APPLICATION OF

SOFT COMPUTING TECHNIQUES IN CIVIL ENGINEERING

APPLICATION OF SOFT COMPUTING TECHNIQUES IN CIVIL ENGINEERING

Edited by

Dr. S.M. Yadav

Professor and Head, Civil Engineering Department
Sardar Vallabhbhai National Institute of Technology, Surat

London • New Delhi

MV Learning
A Viva Books imprint

3, Henrietta Street
London WC2E 8LU
UK

4737/23, Ansari Road,
Daryaganj, New Delhi 110 002
India

ISBN: 978-93-87692-81-7

Printed and bound in India.

Dedicated to

My mother Mrs. Hira and
father Late Mr. Madusudan P. Yadav

CONTENTS

PREFACE

It gives me great pleasure to present this book. This book is an attempt to put together knowledge and experience of soft computing techniques in civil engineering. The principal concern of the book is to show how soft computing techniques can be applied to solve problems in research and practice. An attempt has been made to present various civil engineering research problems using soft computing techniques such as analytic hierarchy process (AHP), fuzzy logic, artificial neural network (ANN), genetic algorithm (GA) and linear programming (LP), etc.

When I got into academics, I was trying all the time to look for reference book which explains the different methods of soft computing used in civil engineering. Students and research scholars always asked for good text or reference books which cover the different methods of soft computing used for civil engineering problems.

My students and research scholars prodded me to present this book. They pointed towards the need for book which has an adequate treatment of analytical tools, like soft computing techniques, in various civil engineering problems. I hope that the book meets these requirements.

Soft computing is a group of techniques and methodologies that can work together to obtain solution in any case. It has a flexible ability and adaptability to situations of the real world. Soft computing techniques are applied to a huge quantity of problems spread in several areas of science.

The field of engineering is a creative one. The problems encountered in this field are generally unstructured and imprecise and tackled by intuitions and past experiences of a designer. The conventional methods of computing, relying on analytical or empirical relations, become time consuming when posed with real-life problems. To study, model and analyse such problems, approximate computer-based soft computing techniques, inspired by the reasoning, intuition, logic and wisdom possessed by human beings, are employed.

In contrast to conventional computing techniques which rely on exact solutions, soft computing aims at exploiting inherent tolerance of imprecision and the trivial and uncertain nature of the problem, to yield an approximate solution in quick time. Soft computing, being a multi-disciplinary field, uses a variety of statistical, probabilistic and optimization tools which complement each other to produce its three main branches viz., Neural Networks, Genetic Algorithms and Fuzzy Logic.

This book can be used as reference book for research scholars and UG/PG students. I wish to thank all the people, without whose contribution, this book would not have come into existence.

Dr. Sanjaykumar M. Yadav

1

Application of AHP Model for Assessing the Performance of Municipal Solid Waste Management

Harshul Parekh[1], Kunwar Yadav[2], S.M. Yadav[3] and Navinchandra Shah[4]

ABSTRACT

The management of municipal solid waste has become an acute problem due to enhanced economic activities and rapid urbanization. This paper aims to demonstrate the application of Analytic Hierarchy Process (AHP) to assign a weight of each performance indicator for Solid Waste Management (SWM). The AHP is a theory of measurement through pairwise comparisons and relies on the judgements of experts to derive priority scales. The degrading state of urban civic services requires an immediate solution to the problems related to mismanagement of urban waste. To evaluate the performance of municipal waste management system forty-four indicators are identified by brainstorming sessions and structured interview, group discussions with experts. A questionnaire was prepared for pairwise comparison of indicators. The judgement of experts through questionnaire is evaluated using AHP model to find out weights of indicators and sub-indicators. For the application of the prepared model, SWM system of two cities as Ujjain and Nashik have been studied and results are discussed. Our research findings will eventually lead to the identification of loopholes in SWM system.

Keywords: *Solid waste management, performance evaluation, analytic hierarchy process, performance indicator*

1. Corresponding Author, Research Scholar, Department of Civil Engineering, S.V. National Indian Institute of Technology, Surat, Gujarat, India, 395007, Email: harshulparekh@gmail.com
2. Associate Professor, Department of Civil Engineering, S.V. National Indian Institute of Technology, Surat, Gujarat, India, 395007, Email: kdy@ced.svnit.ac.in
3. Professor, Department of Civil Engineering, S.V. National Indian Institute of Technology, Surat, Gujarat, India, 395007, Email: smy@ced.svnit.ac.in
4. Director, C.G. Patel Institute of Technology, Gopal Vidyanagar, Bardoli-Mahuva Road, Tarsadi, District Surat, Gujarat, India, 394 350, Email: navin.shah@utu.ac.in

INTRODUCTION

Solid waste management (SWM) is one of the major environmental problems of Indian cities. Economic development, urbanization and improved living standards in cities have increased the quantity and complexity of generated solid waste. Rapid growth in population and urbanization adds greatly to the volume of waste being generated and to the demand for waste management service in municipal areas (Olar, 2003). Problems associated with SWM are acuter in developing country rather than in a developed country. Lack of financial resources and infrastructure to deal with solid waste creates a vicious cycle; lack of resources leads to low quality of service provision which leads to fewer people willing to pay for said services, which in turn further erodes the resource base and so on (Kuniyal, 1998). Another significant factor that contributes to the problem of solid wastes in a developing country scenario is the lack of proper collection and transportation facilities. Improper planning coupled with the rapid growth of population and urbanization serves to add congestion in streets, and as a result, the waste collection vehicles cannot reach such places, thus allowing filth to build up over time. Lack of monetary resources, at times, results in improper or no transportation vehicles for waste disposal adding another dimension to the ever-rising cycle of problems (Jain, 1994; Olar, 2003). Parekh H et al. (2015) studied the identification and assigning weight of indicator influencing performance of municipal solid waste management using AHP for Surat and Ghaziabad.

For cities to be sustainable and to continue their economic development, they must be clean and healthy. In recent years, the government advocates the concept of zero waste in the municipal solid waste and emphasizes the resource recycling to reduce exhaustion of resource (Huang, 2010). This is an important issue to how to reduce the solid waste and effective waste management. It is need of an hour that urban services shall be evaluated and monitored regularly for their performance and outcome in developing countries.

Evaluating and fine-tuning the solid waste system begins with a detailed system analysis. This analysis, or "snapshot," allows selecting the improvement strategies that make the most sense for the community. The system snapshot will also give the baseline needed to develop a cost-effective long-range plan that provides for continuous improvement and reevaluation. In addition, this plan will also supply the flexibility to address the rapid changes occurring within the solid waste management industry.

As a part of monitoring performance of Municipal Solid Waste Management (MSWM) services, it is essential to evaluate their performance. To evaluate the performance of MSW a set of indicators for collection, storage, transportation, treatment and disposal have been designed. Forty-four indicators are identified by brainstorming session and personal discussion with experts of SWM field. These forty-four indicators are classified into eight main area of solid waste management for performance measurement. To find out the weight of indicators, judgement of experts through questionnaire is evaluated using "Analytic Hierarchy Process" (AHP) model. Weight is assigned to indicators and sub-indicators depending on their importance, reliability etc.

PERFORMANCE MEASUREMENT

Performance Evaluation Framework

Performance is the process of regularly measuring the outputs and outcomes of the system. Identification of key performance indicators of the system is the important aspect of the system. Performance measurement enables us to track the amount of work done and the impact of work done on the community. Quantitative and qualitative performance indicators of the municipal waste management system were identified through literature study, brainstorming session and structured interviews with experts. Mode of measurement of indicators was also defined at this stage.

Importance of Performance Measurement in Public Services

Research in both the public and the private sectors demonstrates that "what gets measured gets done" (Training Institute and Data Exchange, 1995) and shows that good systems of performance measurement can facilitate dramatic increases in the quality of services. For managers in the public sector, comparative performance information and results of performance against targets are essential for assessing whether the best services are being provided or purchased and whether those services are meeting the needs of the community. Performance measurement has been gaining importance and is now being applied to programme and service delivery levels (PricewaterhouseCoopers Jan., 2000). Performance measurement results and processes can demonstrate how service activities can make a difference in the community; it helps determine what is working and what needs improvement, and also assists in setting measurable goals and gauge progress against these goals to support sustainability/fundraising efforts.

In the information technology era where data are easily accessible, service providers are facing pressure to evaluate service quality; improve standards of performance to justify solid waste management relative to its cost. Performance measures and indicators allow the *effectiveness* and *efficiency* of a service to be monitored and compared with similar services elsewhere or at an earlier time (METAP, 2004).

Performance Indicators (PI) for Solid Waste Management are useful tools for decision makers, managers and technicians dealing with complex situations which involve deciding, planning or acting (Ljunggren, 2000). It aims to evaluate different alternatives that prove to be intelligible and consistent and also to evaluate the environmental, economic and social consequences of its implementation. Afonso (2009) have also introduced a framework developed for the waste management sector supported by performance indicators. This framework included 167 performance indicators divided into two types of indicators namely context indicators and operational performance indicators.

The framework for performance indicators requires the contribution of each area of SWM. This active participation comprises data collection and treatment and the analysis of performance indicators included in Performance Indicator Framework (PIF). All the elements of PIF should have an adequate measuring procedure that ensures a uniform data collection and also a performance assessment process based on the clear definition and common language. (Coelho and Alegre, 1999). The performance of a solid waste management system is a function of the amount and quality of resources allocated to carry out the services, as well as on the socio-economic development and physical characteristics of the service area. UNEP has proposed performance indicators for solid waste management as shown in Fig. 1.

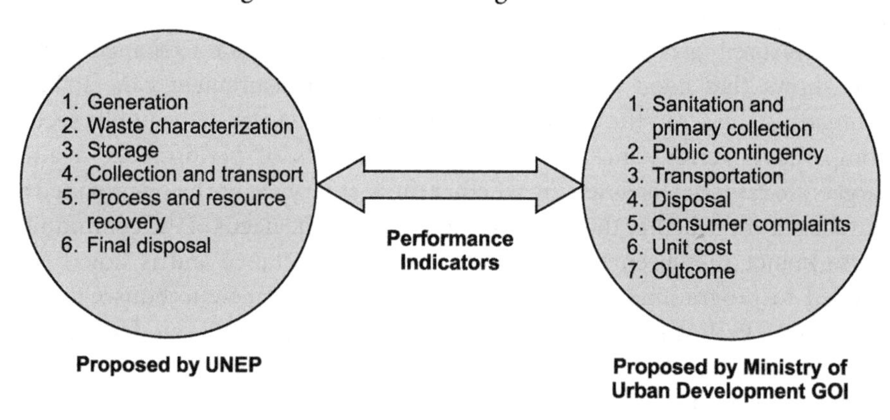

Fig.1. Proposed performance indicators

Ministry of Urban Development has also identified certain performance indicators for service level benchmarking of municipal solid waste management services as shown in Fig. 1.

Identification of key performance indicator is the most important aspect of performance evaluation system. This stage involves identification of all the factors responsible for the various components of WMS (Coelho and Alegre, 1999). The purpose of this indicator is to encourage and assist the urban local body and implementing agency in understanding and appreciating need for:

- Good solid waste management system,
- Regular assessment of their performance to determine and address strength and weakness and
- Ways of measurement of performance.

Structured interviews were conducted with an expert from the solid waste management related field. The outcome of the discussion with experts was listed and used for framing list of indicators for performance evaluation of solid waste services for the present study.

Brainstorming sessions with stakeholders of the solid waste management system were conducted. After studying all the aspects of SWM system, final comprehensive list of forty-five indicators is prepared. These forty-five indicators are classified into eight main area of solid waste management for performance measurement as under.

1. Coverage
2. Transportation
3. Disposal
4. Consumer's complaint
5. Unit cost
6. Outcome
7. Segregation, Recover, Recycle
8. Environmental factors

ANALYTIC HIERARCHY PROCESS

The Analytic Hierarchy Process (AHP) was developed by Saaty (1980) and is sometimes referred to as a Saaty method. It allows decomposition of the problem into hierarchical structures and utilizes its qualitative and quantitative aspects of evaluation. AHP has been widely used in many fields and applied to different types of problems. AHP is considered to be the fundamental approach to decision making. AHP is a multi-criteria decision model that allows both objective and subjective analysis. It is considered to be a comprehensive method that can handle a number of criteria and sub-criteria. In this method, the pair-wise comparison is carried out by the decision maker to select better alternatives. The AHP takes care of inconsistency in judgement and provides methods to improve such inconsistency. Keeney (1982) had divided decision analysis into four phases: (1) structuring the decision problem, (2) assessing the possible impact of each alternative, (3) determining the preference of decision makers, and (4) evaluation and comparison of decision alternatives. This is similar to Saaty's model. Thus, the Analytic Hierarchy Process is suitable to be applied in solid waste management because it can help in decision making.

Strengths and Limitations of AHP

The AHP aids in ranking choices in order of their effectiveness when objectives are conflicting (Coyle, 2004). It incorporates both qualitative and quantitative criteria. It also allows numerous levels of criteria. However, it requires careful selection of experts. With more levels and more criteria, it is difficult to get responses, particularly from respondents in the top of the hierarchy. AHP method also has the possible shortcoming of subjectivity and inconsistency (Hermans, et al., 2008).

Sarkis and Surrandaj (2002) argue that AHP is better than other MCDMs because it has the additional power of being able to combine quantitative and qualitative factors into the decision making process. AHP uses a hierarchical structure of relevant factors that are universally applicable to compose virtually all complex systems, and it is considered a natural problem-solving paradigm in complex situations. Judgement elicitations are derived using a decomposition approach, thereby reducing decision-making errors as observed in experimental studies. AHP has been proven from the decision makers' perspective as well as by recent empirical studies. It also helps multiple stakeholders to arrive at an agreeable solution owing to its structure, and when implemented appropriately, it can build effective consensus.

Michela, et al., (2008) postulates that the advantages of AHP over other methods are due to its compensatory decision methodology that compensates performance of one or more objectives. AHP also allows the application of data, experience, insight, and intuition in a logical hierarchy as a whole. AHP as a weighing method enables decision-makers to derive weights as compared to arbitrarily assigning weight.

Use of AHP in Solid Waste Management

AHP is useful in weighing and decision-making problems and researchers had used it to solve multicriteria decision problems. Dimasalang (1994) determined and applied AHP to select a process of disposing of municipal solid waste for the city of Springfield, Ohio. A multi-criteria decision-making model was developed for ranking alternative for sitting of the municipal solid waste facility.

Sumathi et al. (2008) developed a GIS-based system for optimized landfill site selection for municipal solid waste and AHP was employed to assign weights to different criteria based on expert opinion. Using the comparison matrix among the choices and the information on the ranking of the criteria, AHP generated an overall ranking of the sites.

Qdais (2010) studied landfill leachate treatment options using an analytic hierarchy process to help decision makers prioritize and select leachate management options. Expert choice software was used in the development of the AHP model. A four-level AHP model, with three criteria, nine sub-criteria and five management alternatives were constructed. The benefits of the relative importance scale to indicate the relative strengths of the evaluation criteria were demonstrated in his study.

Mohd et al. (2010) evaluated solid waste treatment technology using AHP, where the multi-level hierarchical structure consisted of objective, criteria, sub-criteria, and alternatives. This was applied for the selection of appropriate solid waste treatment technology. The choices were compared to obtain the weights of the importance of the decision criteria, and the relative performance measures of the various options.

Development of Questionnaire

The questionnaire is prepared with a goal of performance evaluation of municipal solid waste management services. Questions are specific and related to each area of SWM services. The questionnaire is divided into eight main criteria and forty-two sub-criteria as shown in Fig. 3. The relative importance of each main criterion is compared with other seven criteria and similarly, each sub-criteria is compared with other sub-criteria of each group (main area). Respondent has to choose criteria of relative importance and, then decide score depending upon its importance equally, moderate, strong, very strong and extreme importance.

Municipal corporations/municipalities need to effectively implement MSW rule 2000, i.e. for collection, transportation, treatment and disposal of municipal solid waste from the city. Officers, engineers from the solid waste department, consultants, and contractors are chosen as respondents of this questionnaire to give their inputs for importance ranking by their experience. The type of respondents and experience varied from the technical advisor of Central Public Health and Environmental Engineering Organization (CPHEEO), Consultants in private organizations, assistant engineers, deputy commissioner of Municipal Corporation in government organizations, experts of Indian Institute of Technology, National Institute of Technology and Research and Development Laboratories.

The questionnaire was sent to experts associated with solid waste management such as Ministry of Urban Development (GoI), the staff of municipality's consultants, researchers, Academician, solid waste contractor, socialists, and environmentalist for their expert opinion on the relative importance of indicators. The questionnaire was sent to fifty persons from the above-mentioned organizations in hard copy and by email.

Data Analysis using AHP

The Analytic Hierarchy Process basically a tool that permits explicit presentation of evaluation criteria and possibly improves the selection of technology for the solid waste management plan (Mohd and Samah, 2010). The AHP (Saaty, 1980; Saaty, 1990) is known as multi-attribute weighting method for decision support (Madu, 1994). AHP is a decision approach designed to aid in the solution of complex multiple criteria problems in a number of application domains (Saaty, 2000). Keeney (1982) has divided decision analysis into four phases: (1) structure the decision problem, (2) assess the possible impact of each alternative, (3) determine the preference of decision-makers, and (4) evaluate and also compare decision alternative. Thus, Analytical Hierarchy Process is suitable to be applied in solid waste management because it can help in making a decision in importance ranking of criteria more effectively.

Responses received from respondents are analysed using AHP. All the respondents have provided their input of pairwise comparison of criteria measured in Saaty's scale on 1-3-5-7-9. Responses of each respondent, for criteria and sub-criteria, are entered into the matrix, thus nine matrices for each respondent is prepared (one for main criteria and eight for sub-criteria). The geometric mean of all thirty respondents for each category i.e. main criteria and sub-criteria is calculated and one master sheet having one matrix for main criteria and eight matrices for sub-criteria is prepared. This matrix is solved using steps 1 to 8 mentioned in the following paragraph.

Application of AHP for Development Evaluation Model

AHP consists of three main principles, including hierarchy framework, priority analysis and consistency verification. Formulating the decision problem in the form of hierarchy framework is the first step of AHP, with the top level representing overall objectives or goal, the middle levels representing criteria and sub-criteria, and the decision alternatives at the lowest level. Once a hierarchy framework is constructed, a pairwise comparison matrix is set up at each hierarchy and compares each other by using a scale pairwise (Saaty, 1980).

Finally, in the synthesis of priority stage, each comparison matrix is then solved by an eigenvector method to determine the criteria importance and alternative performance. These principles can be elaborated by structuring them in a more encompassing eight steps process as shown in Figure 2.

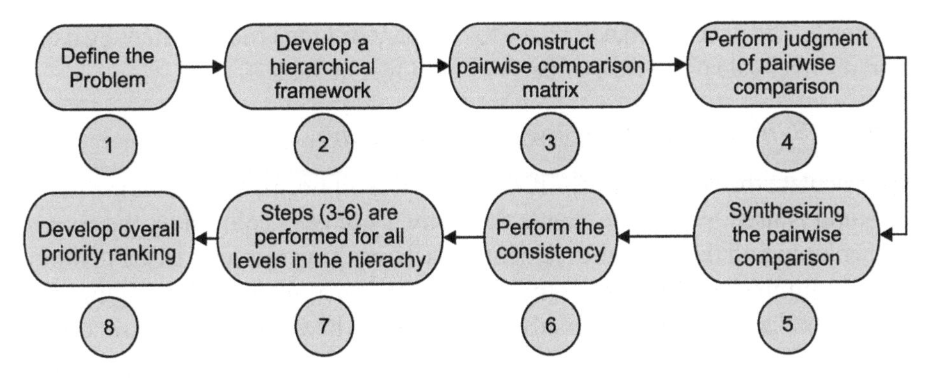

Fig. 2. Steps of AHP

Step 1: Identification of Goal

Evaluation of Solid waste management system is involving multiple factors. The decision problem may involve social, political, technical, and economic factors. Development of an independent model for performance evaluation of Solid Waste Management system is the ultimate goal of the research.

Step 2: Development of Hierarchy Model

In this section, a hierarchy model for evaluation of solid waste management is introduced using AHP.

Step 3: Construction of Pairwise Comparison Matrix

One of the major strengths of AHP is the use of pair-wise comparison to derive accurate ratio scale priorities. A pairwise comparison matrix (size n x n) is constructed for the lower levels with one matrix in the level immediately above. The pair-wise comparisons generate a matrix of relative rankings for each level of the hierarchy. The number of matrices depends on the number of elements at each level. The order of the matrix at each level depends on the number of elements at the lower level that it links to. Fifty-six matrices were constructed for the present study. Table 1 shows the geometric mean matrix of main criteria. Whereas Table 3 shows main criteria of normalized matrix.

The geometric mean of all thirty responses in form of pairwise comparisons is calculated and entered into the one matrix and tabulated in Table 1. Matrix is solved using procedure mentioned in steps 1-8 in the following paragraph.

Step 4: Perform Judgement of Pairwise Comparison

The pairwise comparison begins with comparing the relative performance of two selected items. There are judgements required to develop the set of matrix in step 3. The decision makers have to compare or judge each element by using the relative scale pairwise comparison as shown in Table 1.

Step 5: Synthesizing the Pairwise Comparison

To calculate the vectors of priorities, the average of normalized column (ANC) method is used. ANC is to divide the elements of each column by the sum of the column and then add the element in each resulting row and divide this sum by the number of elements in the row (n). This is a process of averaging over the normalized columns. In mathematical form, the vector of priorities can be calculated as

$$\frac{1}{n}\sum_{j=1}^{n}\frac{aij}{\Sigma_1^n aij}, i, j = 1, 2 \ldots n \tag{1}$$

Table 1 represents one of such matrix.

Table 1. Geometric Mean Matrix of Main Criteria

	Coverage	Transportation	Disposal	Consumers' complaints	Unit cost	Outcome	Segregation, Recover, Recycle	Environmental aspects
Coverage	1.00	2.49	1.18	2.18	2.34	1.79	1.02	0.84
Transportation	0.40	1.00	1.60	2.32	1.26	1.20	0.79	0.98
Disposal	0.84	0.62	1.00	2.03	1.75	7.00	0.59	0.95
Consumers' complaints	0.46	0.43	0.49	1.00	2.07	1.04	0.75	0.95
Unit cost	0.43	0.79	0.57	0.48	1.00	0.89	0.59	0.37
Outcome	0.56	0.83	0.14	0.96	1.13	1.00	0.87	0.56
Segregation, Recover, Recycle	0.98	1.27	1.69	1.33	1.68	1.15	1.00	1.51
Environmental aspect	1.19	1.02	1.05	1.05	2.69	1.79	0.66	1.00
Total of column	5.87	8.456	7.736	11.36	13.93	15.85	6.277	7.158

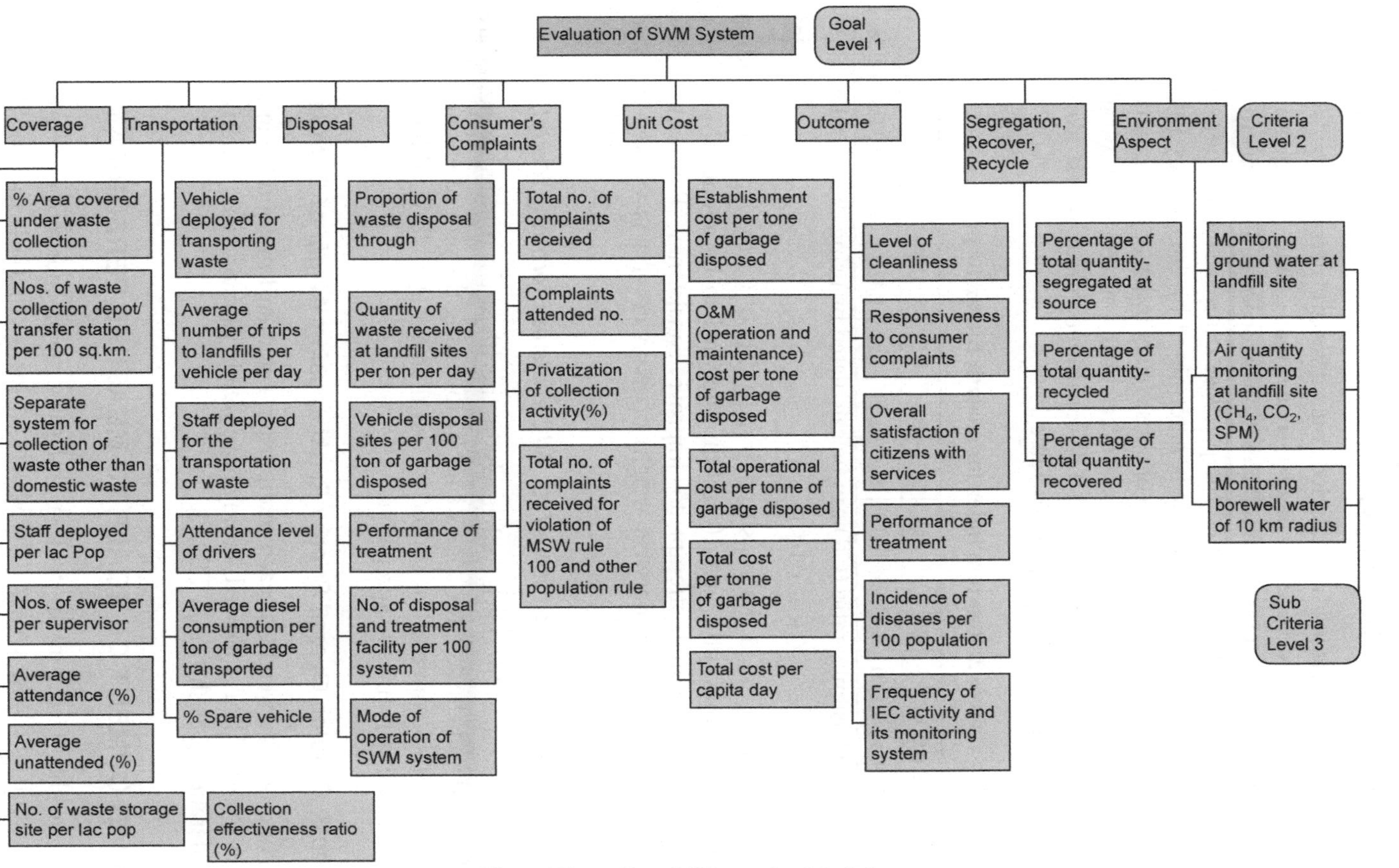

Fig. 3. Three Level Hierarchy Model

Step 6: Perform the Consistency

Since the comparisons are carried out through personal or subjective judgements, some degree of inconsistency may occur. To guarantee the judgements are consistent, the final operation incorporated in order to measure the degree of consistency among the pairwise comparison by computing the consistency ratio. The consistency is determined by consistency ratio (CR). Consistency ratio (CR) is the ratio of consistency index (CI) to random index (RI) for the same order of matrix. To calculate the consistency ratio (CR) computation for eigenvalue (λ_{max}) and consistency index is first worked out. The Random Index (RI) as given in Table 2.

Table 2. Random index for analytic hierarchy process

Size of Matrix (n)	1	2	3	4	5	6	7	8	9	10	11	12
Random Index (RI)	0	0	0.58	0.9	1.12	1.24	1.32	1.41	1.45	1.49	1.51	1.58

Then consistency ratio (CR) can be calculated by using the formula

$$CR = CI/RI \tag{2}$$

If the ratio (called the consistency ratio CR) of CI to that from random matrices is significantly small (carefully specified to be about 10% or less). Accept the estimated attempt to improve consistency. Following above procedure weight or otherwise weight of main criteria is worked out and presented in Table 2.

Step 7: Step 3-6 are Performed for All Levels in Hierarchy Model

Each criterion, sub-criteria (all levels in the hierarchy model) weights are calculated and checked for consistency by following the procedure mentioned in Steps 3 to 6. Value of CR should be less than 0.1 at all levels then and only then the judgements are acceptable.

Step 8: Develop Overall Priority Ranking using Weight of Criteria

To find out the weight of each main criteria standard method for analysis of data as per AHP is followed. From above weight of main criteria derived after 6th iteration is as shown in Table 3.

DEVELOPMENT OF SCALE TO MEASURE PERFORMANCE

The weight of each main criteria and sub-criteria indicators are useful for evaluation of SWM system. The weight of main criteria is calculated by the

Table 3. Main criteria of normalized matrix

	Coverage	Transportation	Disposal	Consumers' Complaints	Unit cost	Outcome	Segregation, Recover, Recycle	Environmental aspects	Weight after 1st iteration	Weight after 6th iteration
Coverage	0.17	0.294	0.153	0.192	0.168	0.113	0.162	0.117	17%	**17.19%**
Transportation	0.069	0.118	0.207	0.205	0.091	0.076	0.126	0.137	13%	**12.92%**
Disposal	0.144	0.074	0.129	0.179	0.125	0.442	0.094	0.133	16%	**15.00%**
Consumers' complaints	0.078	0.051	0.064	0.088	0.149	0.066	0.12	0.133	9%	**9.18%**
Unit cost	0.073	0.094	0.074	0.043	0.072	0.056	0.095	0.052	7%	**7.20%**
Outcome	0.095	0.099	0.018	0.085	0.081	0.063	0.139	0.078	8%	**8.41%**
Segregation, Recover, Recycle	0.168	0.15	0.218	0.117	0.121	0.072	0.159	0.211	15%	**16.16%**
Environmental aspect	0.203	0.121	0.136	0.093	0.193	0.113	0.106	0.14	14%	**13.94%**

method described above. A similar procedure is followed for each set of sub-criteria and weight of sub-criteria is calculated as follows:

The weight of sub-criteria in each set is multiplied by the weight of main of its own criteria to determine its weight in the whole system. e.g. weight of coverage is 17.19/100, now weight of % area covered under waste collection system is 14.79% hence the final weight of % area covered under waste collection system is 14.79% × 17.19 = 2.54/100. The weight for all forty-four parameters is derived using the same process as described here.

Now, to evaluate the performance of each indicator of SWM system some measurement index is required. SWM data of cities fall in the very wide range. Now the question is what score should be given to a particular data indicator? It can be explained by the following example for sub-criteria of % area covered waste collection system for Ujjain and Nashik.

Name of city	Performance of city in % area covered waste collection system
Ujjain	100%
Nashik	100%

The weight of sub-criteria of "%" area covered waste collection system is 2.54. For both cities data for "%" area covered waste collection system fall in the range of 91-100%. So, the score for both the cities is 2.54 i.e. 100% score.

As discussed earlier, Solid Waste Management system of any city is the very complicated system. It involves various operations and a large number of manpower. It is not only limited to collection, transportation and treatment of solid waste but minute factors also play a vital role in determining the efficiency of the entire system. A questionnaire was prepared with possible performance range for each indicator and it was sent to 140 experts. The range of scale in form of maximum and minimum service level is decided by inputs from thirty-five respondent experts. Input data in form of minimum and maximum service level has been analysed to frame range of performance indicator. Sample index for a range of performance level is described in Table 4. A similar procedure is followed for calculation of the score for all forty-four parameters.

Table 4. Sample index for range of data for various indicators

Indicator	Weight of Sub-criteria	Score for range of performances				
		20%	40%	60%	80%	100%
% area covered waste collection system	2.54	51-60	61-70	71-80	81-90	91-100
Total number of complaints received per day per one lac population	1.58	>11	9-11	6-8	3-5	≤2

PERFORMANCE EVALUATION OF SWM SYSTEM OF UJJAIN AND NASHIK

Ujjain is an ancient city of central India in the Malwa region of Madhya Pradesh. Ujjain stands glamorously among many other Indian sacred and holy cities. The permanent inhabitant population of the city as per Census 2001 is 4,30,427 inhabiting an area of 180 sq.km. As per the latest Census - 2011, the city has a population of 5,15,215 with a growth rate of 19.70%.The Ujjain Municipal Corporation estimates that daily 225 tonnes of solid waste are generated in the urban area. Ujjain is a city of great religious significance, having the sudden rise of floating population which further increases the issue of solid waste management.

The city of Nashik is situated in the State of Maharashtra, in the northwest of Maharashtra, on 19 deg. N 73 deg. E coordinates. It is connected by road to Mumbai (185 km) and to Pune (220 kms). The urban agglomeration (UA) population has increased from 10.72 lakhs in 2001 to 15.84 lakhs in 2011geographical proximity to Mumbai (economic capital of India) and forming the golden triangle with Mumbai and Pune has accelerated its growth. The development of the past two decades has completely transformed this traditional pilgrimage centre into a vibrant modern city. In order to cope with massive problems that have emerged as a result of rapid urban growth, the Government of India, Ministry of Urban Development has selected cities under the Jawaharlal Nehru National Urban Renewal Mission (JNNURM) launched in 2005-06. Nashik is one of the selected JNNURM cities.

In order to evaluate the performance of solid waste management system, data related to SWM services of Ujjain and Nashik is collected from concerned authority by author and analysed using the model. Performance evaluation of SWM systems of Ujjain and Nashik is carried and results are discussed in preceding paragraphs.

Coverage

Coverage of SWM Services is related to waste collection, door-to-door garbage collection, street sweeping, nos. of community bins, manpower for the collection of waste & their sufficiency, the efficiency of waste collection system in the city. The score of main criteria coverage is presented in Table 5.

It is observed from the above evaluation that both the cities have 100% score in area coverage for waste collection. Nashik has a better system for collection of all different types of waste which is really appreciable while Ujjain needs to improve for a separate collection system for all different kind of waste. It is observed that all type of solid waste generated is mixed with municipal

Table 5. Score of main criteria coverage

Code	Coverage	Weight of sub-criteria	Score of UMC	Score of NMC
C11	% area covered under waste collection	25.4	25.4	25.4
C12	Separate system for collection of waste other than domestic waste (biomedical waste, C&D waste, hotel waste, market waste, dead animal disposal)	18.7	7.5	18.7
C13	Staff deployed per lakh population	11.4	11.4	9.90
C14	No. of sweeper per supervisor	14.6	7.20	6.50
C15	Average attendance (%)	22.6	18.1	20.30
C16	Beans left unattended (%)	20.8	10.4	8.30
C17	No. of waste storage site per lakh population	28.0	22.4	6.80
C18	Collection effectiveness ratio (%)	30.4	13.0	30.4
		171.9	115.4	126.3

solid waste including biomedical waste, demolition waste, highly putrescible waste, slaughterhouse waste, dead animal and inert waste. Nashik scores well in almost all criteria except staff deployed per lakh population and no. of waste storage site per lakh population.

Transportation

Transportation is the second important stage of the solid waste management system. This being a very important element as it involves a large proportion of the capital and operating cost and cause an impact on both, primary collection and processing. Transportation stage can be defined as transportation of stored solid waste from various places in specially designed vehicles to the disposal site in the most hygienic way. The objective of the system is the transportation of collected refuse from specific collection points to the disposal site at minimum cost and removal of the waste at regular intervals from all collection points. The score of transportation is presented in Table 6.

In Ujjain, there is a strong decline in the standard of services with respect to collection, transportation and disposal. Household waste is disposed of either via door-to-door collection or in the primary collection points or dustbins provided by Municipal Corporation. In Ujjain generally waste is transported and simply dumped in an abandoned area whereas Nashik corporation has made efforts to improve the solid waste management and the concept of a unique system of the house to house garbage collection by the private agencies.

Table 6. Score of transportation

Code	Transportation	Weight of sub-criteria	Score of UMC	Score of NMC
C21	Vehicles deployed for transporting waste per 100 tonne	12.2	10.5	1.7
C22	Average number of trips to landfills per vehicle per day	16.4	12.3	4.1
C23	Efficiency in carrying garbage = garbage carried/rated capacity of vehicle	22.5	10.2	14.8
C24	Staff deployed for transportation of waste/ 100 tonne	10.9	1.6	2.1
C25	Attendance level of drivers	18.5	13.9	13.9
C26	Average trips per driver per day	21.7	16.3	8.1
C27	Average diesel consumption per tonne per km of garbage transported	13.7	0.00	0.00
C28	% of spare vehicle	13.2	1.3	4.8
		129.2	66.1	49.5

Ujjain has better performance than Nashik in almost all criteria. Where Nashik has performed better in efficiency in carrying garbage.

Disposal

The final stage of SWM is disposal. Indicators which are used to monitor the amount of solid waste that is being disposed of in many ways that have different implications in environment and conservation of resources. Waste disposed of through composting, waste-to-energy, sanitary landfilling, open dumping with manpower and machinery are monitored. The score of disposal is presented in Table 7.

Disposal of solid waste is very important and sensitive criteria. Evaluation of disposal criteria reveals the poor performance of both cities in the scientific disposal of solid waste. While both cities have good performance for no. of disposal and treatment facility per 100 sq.km.

Consumer's Complaint

This indicator is used to monitor the number of complaints being received with respect to the population of the city, time is taken to attend complaints, IEC

Table 7. Score of disposal

Code	Disposal	Weight of sub-criteria	Score of UMC	Score of NMC
C31	Proportion of waste disposed of through scientific disposal manner	21.8	0	0.3
C32	Percentage of waste received at landfill sites per day	19.9	17.4	0
C33	Staff at disposal sites per 100 tonnes of garbage disposed	11.4	1.10	4.0
C34	Vehicles at disposal sites per 100 tonnes of garbage disposed	11.5	0	2.4
C35	Performance of the treatment (compost plant, west to energy, RDF plant, eco bricks)	36.1	22.1	12.8
C36	No. of disposal and treatment facility per 100 sq.km	23.0	23.0	23.0
C37	Mode of operation of SWM system	26.4	13.2	26.4
		150.0	76.8	68.9

activity performed by SWM department. The score of consumer's complaint is presented in Table 8.

Table 8. Score of consumer's complaint

Code	Consumer's complaint	Weight of sub-criteria	Score of UMC	Score of NMC
C41	Total number of complaints received per day per one lakh population	15.8	14.2	12.6
C42	Percentage of complaints attended	29.0	23.2	23.2
C43	Privatization of collection activity (%)	16.4	0	13.1
C44	Total number of complaints received for violation of MSW rule 2000 and other pollution norms	30.5	23.7	30.5
		91.8	61.1	79.4

Consumer satisfaction is reflected in the status of consumer complaints. From the evaluation of consumer's complaint criteria; suggests overall good performance for both the cities. Nashik has better performance compared to Ujjain in a total number of complaints received for violation of MSW rule 2000 and other pollution norms.

Unit Cost

This indicator includes expenditure related to the annual establishment cost, O&M cost for the quantity of garbage disposed of. Revenue from recyclable material, the sale of compost, RDF, energy etc. The score of unit cost is presented in Table 9.

Table 9. Score of unit cost

Code	Unit cost	Weight of sub-criteria	Score of UMC	Score of NMC
C51	Establishment cost per tonne of garbage disposed	8.1	3.8	7.6
C52	O&M (operation and maintenance) cost per tonne of garbage disposed	9.8	1.8	7.8
C53	Total operational cost per tonne of garbage disposed	14.2	2.8	12.6
C54	Total cost per tonne of garbage disposed	16.5	3.0	16.5
C55	Total cost per capita day	23.5	16.7	23.5
		72.0	28.1	68

Cost again plays an important role in any system. Financial evaluation of solid waste system shows where the money is being spent. It also reveals to what extent it is fruitful. Evaluation of unit cost criteria shows good performance for Total operational cost per tonne of garbage disposed of in both cities. Establishment cost per tonne of garbage disposed and total cost per tonne of garbage disposed of indicates that in case of UMC amount spent is very high compared to NMC.

Outcome

Indicators for the outcome of SWM services i.e. level of cleanliness, the satisfaction level of citizen and responsiveness to consumer's complaint measured by a consumer survey.

The score of outcome is presented in Table 10.

The outcome shows overall impact of SWM system among the citizens. Both cities score quite a good score in all parameters except frequency of IEC activity and its monitoring system. Evaluations show that concerned authorities should aware people through various media available and educate citizens. It also indicated that Nashik has a better complaint redressal system than Ujjain.

Table 10. Score of outcome

Code	Outcome	Weight of sub-criteria	Score of UMC	Score of NMC
C61	Level of cleanliness	20.7	6.9	13.8
C62	Responsiveness to consumer complaints	11.5	4.0	12.0
C63	Overall satisfaction of citizens with services	18.4	6.0	12.0
C64	Incidences of diseases per 1000 population	18.4	0	0
C65	Frequency of IEC activity and its monitoring system 1. Awareness/Education School programmer 2. Print media (handbill, hoardings) 3. Electronic media (TV, radio) 4. Street play 5. Door-to-door awareness campaign	15.2	6.3	1.30
		84.1	23.2	39.1

Segregation, Recover and Recycle

The success of SWM services mainly depends on awareness of citizen for segregation of waste source and on city administration, it is a responsibility to recycle, recover (material energy) from segregated waste. The score of segregation, recover and recycle is presented in Table 11.

Table 11. Score of segregation, recover and recycle

Code	Outcome	Weight of sub-criteria	Score of UMC	Score of NMC
C71	Percentage of total quantity – Segregated at source	69.5	0.0	10
C72	Percentage of total quantity – Segregation at treatment plant	24.4	4.0	4.0
C73	Percentage of total quantity – Recycled	35.3	0.0	0.0
C74	Percentage of total quantity – Recovered	32.4	0.0	0.0
		161.6	4.0	14.0

Evaluation of segregation, recover and recycle clearly shows a present scenario in both the cities. Ujjain has no quantity of waste is segregated at source. At treatment plant also segregation is not up to the mark. Recycling and recovery of waste are absent in both cities. This situation can be overcome by educating people. Segregation at source should be motivated. Authority can distribute separate dustbins for dry and wet waste. All other three parameters will improve eventually as they are directly connected to segregation at source.

Environmental Aspects

Regular monitoring of groundwater depth, quality of water in borewell near landfill site, results will show contamination of groundwater due to leachate percolation, air quality monitoring at landfill site are important parameters.

The score of environmental aspects presented in Table 12.

Table 12. Score of environmental aspects

Code	Environmental aspects	Weight of sub-criteria	Score of UMC	Score of NMC
C81	Monitoring ground water at landfill site	44.0	0.0	44.0
C82	Monitoring ground water within 5 km radius	51.0	0.0	51.0
C83	Air quality monitoring at landfill site (methane, CO_2, SPM)	44.4	0.0	44.4
		139.4	0.0	139.4

Evaluation of environmental aspects shows that monitoring of groundwater, air quality monitoring at landfill site within 5 km radius is absent at Ujjain. It should be performed to conserve our environment in a better way. Nashik has better performance in all environmental aspects.

Table 13. Summary of performance of UMC and NMC for SWM system

Code	Variables	Weight of main criteria	Final score UMC	Final score NMC
C1	Coverage	171.9	115.4	126.3
C2	Transportation	129.2	66.1	49.5
C3	Disposal	150.0	76.8	68.9
C4	Consumers' complaints	91.8	61.1	79.4
C5	Unit cost	72.0	28.1	68.0
C6	Outcome	84.1	23.2	39.1
C7	Segregation, recover, recycle	161.6	4.0	14.0
C8	Environmental aspect	139.4	0.0	139.4
		1000.0	374.7	584.6

Table 13 clearly shows that Nashik has better performance in coverage, consumers' complaints, outcome, unit cost, segregation, recover, recycle and environmental aspects. Ujjain has better performance in disposal and transportation. Both cities perform poor in segregation, recover and recycle. Average performance is in outcome.

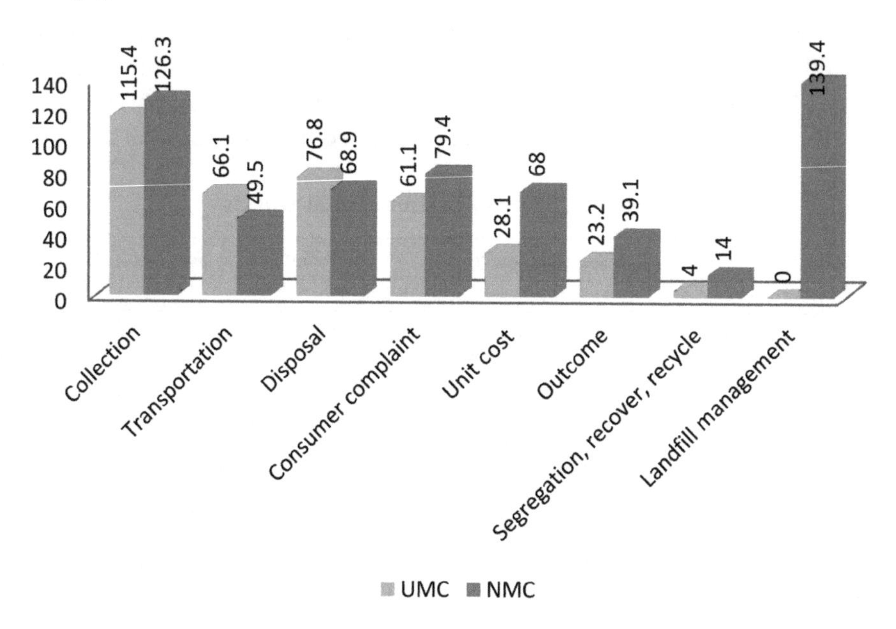

Fig. 4. Overall performance of Ujjain and Nashik in main criteria

CONCLUSION

The municipal solid waste management system is a multi-dimensional activity which involves machines, manpower and participation of peoples of the city. To evaluate the performance of municipal waste management system indicators are identified by brainstorming sessions and personal discussion with experts. A questionnaire was prepared for pairwise comparison of indicators. The judgement of experts through questionnaire is evaluated using "Analytic Hierarchy Process" (AHP model) to find out weights of indicators.

It is observed that performance of some indicators is influenced by the performance of other indicators like the cost of transportation does not only depend on manpower, machinery, spare vehicles but also depends on distance to the landfill site, mode of operation i.e. departmental, contractual or Public Private Partnership (PPP) mode.

The present study reveals that performance of Urban Local Bodies (ULBs) is quite good in areas of collection and transportation. But areas like segregation, recover, recycle and environmental aspects need more attention. Public participation should be encouraged for segregation at source.

Present model for evaluation of Municipal Solid Waste management system is simple linear model, it does not account for influence of other indicators within this model and external factors like staff deployed per lakh population is influenced by % area covered under waste collection, mode of operation, use of machinery in SWM no. of waste collection depot is influenced by implementation of door-to-door garbage collection system.

REFERENCES

Afonso, C. (2009). Municipal Solid Waste Performance Indicators, National Laboratory of Civil Engineering (LNEC).

Coelho, S.T. and Alegre, H. (1999). Performance Indicators for Water and Sanitation Systems, Hydraulic Technical Information 40, Laboratório Nacional de Engenharia Civil, Lisbon, Portugal.

Huang, Y.S. (2010). "Evaluation of Solid Waste Management Strategies in the Taipei Metropolitan Area of Taiwan." *Air & Waste Manage. Association*, 56(5), pp. 650-656.

Jain, A. (1994). "Solid Waste Management in India." 20th WEDC Conference, Affordable Water Supply and Sanitation, Colombo, Sri Lanka, pp. 177-182.

Keeney, R.L. (1982). "Decision Analysis: An Overview." *Operations Research*, 30(5), pp. 803-838.

Kuniyal, J. (1998). "Public Involvement in Solid Waste Management in Himalayan Trails in and around the Valley of Flowers, India." *Resources, Conservation and Recycling*, 24(3-4), pp. 299-322, doi:10.1016/S0921-3449(98)00056-1.

Ljunggren, M. (2000). "Modelling National Solid Waste Management." *Waste Management & Research*, 18(6), pp. 525-537.

Madu, C.N. (1994). "A Quality Confidence Procedure for GDSS Application in Multicriteria Decision Making." *IIE Transactions*, 26(3), pp. 31-39.

Ministry of Urban Development (2000). CPHEEO Manual on Municipal Solid Waste Management, Ministry of Environment and Forest, New Delhi.

Ministry of Urban Development (2010). Documentation of Best Practices, Vol. 3, National Institute of Urban Affairs, New Delhi

Ministry of Urban Development (2010). Handbook on Manual Service Level Benchmarks, Ministry of Urban Development, Government of India, New Delhi.

Ministry of Urban Department (2010). Improving Service Outcomes (2008-09), Service Level Benchmarking Databook, Government of India, New Delhi.

Olar, Z. (2003). Urban Solid Waste Management: Waste Reduction in Developing Nations, Michigan Technological University, Working Paper 2003.

Parekh, H., Yadav, K., Yadav, S., and Shah, N. (2015). Identification and Assigning Weight of Indicator Influencing Performance of Municipal Solid Waste Management using AHP. *KSCE Journal of Civil Engineering*, 19(1), pp. 36-45.

Saaty, T.L. (1980). Analytic *Hierarchy Process, Planning, Priority Setting, Resource Allocation*. New York: McGraw-Hill.

___________. (1990). *Multicriteria Decision Making: The Analytic Hierarchy Process*. RWS Pittsburgh Publication: Pittsburgh.

2

Application of Analytic Hierarchy Process in Measurement of Construction Project Performance

Mrunal Jani[1], Devanshu Pandit[2] and S.M. Yadav

ABSTRACT

Many researchers have held construction industry as an under-performing sector compared to other industries but construction industry is one of the most important industries all over the world. So the performance measurement of this industry is also a big challenge nowadays. This paper presents a framework for performance measurement of the construction projects. The main objective is to measure the performance of the finished building projects. This framework is based on performance index which was generated by applying two rounds of questionnaire survey. First round is for selection of the factors and the second round is to give weightage to the factors which are shortlisted from first round. Based on the survey and interview data collected on performance measurement, project performance index was developed. Factors selected for the performance index are time, cost, quality, client's satisfaction, communication, safety & health and environment. This was implemented to cases and the result shows that this method can be an effective tool for the performance measurement of the finished building projects.

Keywords: *Performance indicators, performance measurement, analytical hierarchical process, performance index*

INTRODUCTION

Indian construction industry is one of the fastest growing construction industries internationally and the second largest employer in India. Construction industry is very complex in nature because it contains large number of people as consultants, contractors, clients, stakeholders and many others. Construction project development also involves numerous parties,

1. Deputy Manager, JMC Projects Pvt. Ltd., Mumbai
2. Professor, CEPT University, Ahmedabad

various approaches, different phases and stages of work (Takim and Akintoye, 2002). So, construction projects in India suffer from many problems and complex issues in performance such as cost, time and safety.

Project performance measurement is defined as the process of evaluating performance of the project relative to defined goal (Azlan, 2010). In the 1960s, project success was measured entirely in technical terms: either the product worked or it did not. In the 1980s, the following definition for project success was offered: project success is stated in terms of meeting three objectives: 1) completed on time, 2) completed within budget and 3) completed at the desired level of quality. All three of these measures are internal to a project and do not necessarily indicate the preferences of the end user or the customer. In the late 1980s, after the introduction of TQM, a project was considered to be a success by not only meeting the internal performance measures of time, cost and technical specifications but also make sure that the project is accepted by the customer; and resulted in customers allowing the contractor to use them as a reference (Khosravi 2011).

Navon (2005) divides the main performance problem into two groups: (a) unrealistic target setting (i.e., planning) and (b) causes originating from the actual construction; and also stated that international construction projects performance is affected by more complex and dynamic factors than domestic projects; frequently being exposed to serious external uncertainties such as political, economical, social, and cultural risks, as well as internal risks from within the project. Long et al. (2004) remarked that performance problems arise in large construction projects due to many reasons such as: incompetent designers/contractors, poor estimation and change management, social and technological issues, site related issues and improper techniques and tools. Also found that the traditional performance measurement systems have problems because of large and complex amount of information with absence of approaches to assist decision maker understand, organize and use such information to manage organizational performance.

PERFORMANCE MEASUREMENT METHODOLOGIES IN CONSTRUCTION

According to Yang (2010), methods for performance measurement in construction industry are gap analysis, integrated performance index, statistical methods, data envelopment analysis (DEA) method. Some of the popular gap analysis-based techniques for performance measurement in construction are the "spider" or "radar" diagrams, and the "Z" chart. These tools are graphical in nature and could be easily understood because they are capable of showing multiple dimensions simultaneously. Rankin (2011) used radar-diagrams to show the cost predictability, time predictability, cost and time per unit to measure the performance of Guyana's construction industry.

Statistical methods, such as regression analysis, multiple regression and various descriptive statistics are used to analyse data in performance measurement. Yeung (2010) also suggested that by using multiple regressions analysis we can not only measure but prediction of the project performance can also be done. Data envelopment analysis adopts the linear programming technique to evaluate the efficiency of the analysed units. DEA is able to evaluate the performance quantitatively as well as qualitatively.

Many researchers have generated performance index for different sectors, Lam et al. (2008) introduced the development of a Project Success Index (PSI) for design-build projects and aims to find out the determinants of success for Design and Build (D&B) projects. Principal Component Analysis was used by Lam for the analysis. Hamilton (1997) also developed a success index to benchmark the project success, the success index not only include project success but also include the operating success. The success index includes the following objective measures: cost performances, schedule performance, plant utilization, and design capacity obtained after six months of operation. Variable weights were constructed using subjective data obtained through 131 interviews. With the combination of objective and subjective measures, an effective benchmark for project success was developed. Pillai et al. (2002) in their studies, proposed a model of performance measurement for R&D projects. In this model, they identify four important areas: the project phases, the performance indicators associated with each phase, the stakeholders and the performance measurements.

Pilliai et al. (2002) derived the integrated performance index from the following three steps:

1. Identification of important phases of project life cycle;
2. Identification of key factors in each phase; and
3. Integration of all the key factors into an integrated performance index.

QUESTIONNAIRE SURVEY (SELECTION OF THE FACTORS)

Questionnaire-1 consists of fourteen performance indicators (selected from literature review) which includes; cost, time, quality, client's satisfaction, safety and health, communication, functionality, innovation, productivity, contractor's satisfaction, architect's satisfaction, environment, design change, working hours. The questionnaire survey conducted in order to assign a weight factor for the selected project success criteria. In this questionnaire, the respondents were asked to identify a significant degree of the selected project success criteria on a five-point Likert scale (1) not important, (2) low important, (3) medium important, (4) highly important, (5) extremely important. Total forty-four persons responded back, among them 58% were contractors, 26% were consultants and 16% were clients. The response rate for the first questionnaire was 80%.

Table 1. Analysis of questionnaire-1

Factors (i)	Frequency of response (ii)					Mean (iii)	Variance (iv)	Std dev. (v)	Co-efficient of variation (vi)	Ranking (vii)
	1	2	3	4	5					
Cost	0	1	6	27	9	4.02	0.45	0.67	0.167	3
Time	0	0	8	14	21	4.30	0.60	0.77	0.180	4
Quality	0	1	3	16	23	4.42	0.53	0.73	0.166	2
Safety and health	1	0	12	16	14	3.79	0.65	0.80	0.212	6
Client's satisfaction	0	1	7	21	14	4.49	0.35	0.59	0.132	1
Communication	0	1	20	19	3	3.49	0.59	0.77	0.220	7
Functionality	0	3	11	24	5	3.72	0.59	0.77	0.206	5
Innovation	1	10	17	9	6	3.21	1.07	1.04	0.323	13
Productivity	1	3	23	13	3	3.40	0.63	0.79	0.233	9
Contractor's satisfaction	2	5	25	11	0	3.05	0.57	0.75	0.248	10
Architect satisfaction	3	14	9	16	1	2.95	1.09	1.05	0.354	14
Environment	0	4	14	22	3	3.53	0.64	0.80	0.226	8
Design change	1	10	20	8	4	3.09	0.90	0.95	0.306	11
Working hours	2	8	18	11	4	3.16	1.00	1.00	0.316	12

Column 1 of Table 1 shows the factors selected for the questionnaire-1, column 3 shows the mean (μ) of the responses, column 4 and 5 shows the variance (σ^2) and standard deviation (σ). By using mean and standard deviation coefficient of variation (c) is defined as the ratio of standard deviation (σ) and means (μ). According to the Lotfi et al. (2010) ranking of the factors can be done by using coefficient of variation, lower the coefficient of variation higher the stability (Table 1).

From the analysis of Table 1 eight factors were shortlisted which are (1) cost, (2) time, (3) quality, (4) safety and health (5) client's satisfaction, (6) communication, (7) functionality and (8) environment. Table 2 shows the correlation between selected factors, which suggest that factors are not highly correlated with each other and some of them are insignificantly correlated with each other.

Table 2. Correlation between selected factors

Correlation matrix	1	2	3	4	5	6	7	8
1	1.00	0.04	-0.075	0.06	0.39**	0.06	0.02	.23
2		1.00	0.20	0.16	-0.04	0.28	0.13	.003
3			1.00	0.27	0.31*	0.06	0.35*	0.25
4				1.00	0.17	-0.03	0.31*	0.20
5					1.00	0.05	0.30	0.43**
6						1.00	0.19	0.16
7							1.00	0.22
8								1.00

(**. Correlation is significant at the 0.01 level)
(*. Correlation is significant at the 0.05 level)

Assigning Weights

From the analysis of the first questionnaire a new questionnaire was developed with the shortlisted factors to assign weights to them. Weights can also be derived from first questionnaire by using relative importance index (RII) or any other method but first questionnaire does not compare the one factor with the other. So if the first questionnaire is used for weighing then the result will not be trustworthy.

There are many methods to assign weights to factors, however in this paper Analytic Hierarchical Process (AHP) is used to assign weights, which is widely used and effective method for weighing. The Analytic Hierarchy Process (AHP) is a decision-making procedure originally developed by Thomas Saaty; its primary use is to offer solutions to decision problems in multivariate environments, in which several alternatives for obtaining given objectives are compared under different criteria. AHP technique involves the following steps.

In the first step, the relative importance of the performance indicator is judged by a pairwise comparison of the indicators. The respondents state their preferences for each pair of the set, considering the importance of one factor against the other by ranking how much time they prefer the factor, using intensity scale for AHP i.e., equally preferred, weekly preferred, moderately preferred, strongly preferred, or absolutely preferred, which would translate into pairwise numerical values of 1, 3, 5, 7 and 9, respectively, with 2, 4, 6, and 8 as intermediate values (as shown in Table 3).

Table 3. Scale for pairwise comparison in AHP

Compare index-A with index-B	The evaluation value of index-A	
By comparison, A is extremely more important than B	9	2, 4, 6, 8 is the mid-value of these adjacent values
By comparison, A is much more important than B	7	
By comparison, A is more important than B	5	
By comparison, A is slightly more important than B	3	
By comparison, A is as important as B	1	1/2, 1/4, 1/6, 1/8 is the mid-value of these adjacent values
By comparison, A is slightly less important than B	1/3	
By comparison, A is less important than B	1/5	
By comparison, A is much less important than B	1/7	
By comparison, A is extremely less important than B	1/9	

The judgement matrix is synthesized to calculate the priority vector; it is commonly performed by one of the mathematical techniques, such as eigenvectors, mean transformation, or geometric mean. In this case, geometric mean technique has been applied, which is calculated as n^{th} root of product of n numbers.

Finally, the level of inconsistency in the judgement matrix is checked by computing the consistency ratio (CR) as

$$CR = (CI/RCI) \times 100 \quad (1)$$

Where, CI = consistency index (equation-2) and

RCI = random consistency index (equation-3)

$$CI = (\lambda_{max} - n) / (n - 1) \quad (2)$$

$$RCI = (\lambda'_{max} - n) / (n - 1) \quad (3)$$

Where λ_{max} = maximum eigen value of the judgement matrix;

n = dimension of the pairwise comparison of judgement matrix;

λ'_{max} = average eigen value of the judgement matrix derived from randomly generated reciprocal matrices using the scale (1/9, 1/8,1/7, 1/6, 1/5, 1/4, 1/3, 1/2, 1, 2, 3, 4, 5, 6, 7, 8, 9)

Table 4 presents the RCI values used to check the inconsistency in the AHP judgement matrix.

Table 4. Average random consistency index table

Matrix size	1	2	3	4	5	6	7	8	9	10
RCI	0	0	0.58	0.90	1.12	1.24	1.32	1.41	1.45	1.49

The standard rule recommended by Saaty indicates that the Consistency Ratio should be less than or equal to 10% to be consistent in pairwise judgements. Consistency ratio has to be checked for the final pairwise judgement, which is calculated from geometric mean. In this case consistency ratio is 1%. From that matrix weightage of each key performance indicator was calculated by using analytic hierarchical process which is shown in Table 5.

Table 5. Weightage of each factor

KPI	Weightage
Client's satisfaction	9.51
Time	12.83
Quality	17.02
Cost	11.78
Safety and health	16.64
Functionality	11.73
Communication	7.74
Environment	12.75

Generation of Project Performance Index (PPI)

The general form of project performance index can be expressed as follows:

$$PPI = f\,(CS, T, Q, C, SH, F, Co, E) \tag{4}$$

Where, PPI = project performance index, f = function, CS = client's satisfaction, T = time, Q = quality, C = cost, SH = safety and health, F = functionality, Co = communication, E = environment.

Project performance index (PPI) is a linear combination of weighted parameters. General equation of performance index can be written as:

$$PPI = \sum_{i=1}^{n} W_i P_i \tag{5}$$

Where, P_i = value of parameter "i"

W_i = weight of parameter "i"

i = an integer value identifying one of the key performance indicator

Project performance index can be computed as follows,

$$\begin{aligned} PPI &= (W_{CS} \times P_{CS}) + (W_T \times P_T) + (W_Q \times P_Q) + (W_C \times P_C) + \\ &\quad (W_{SH} \times P_{SH}) + (W_F \times P_F) + (W_{Co} \times P_{Co}) + (W_E \times P_E) \\ &= (9.51 \times P_{CS}) + (12.83 \times P_T) + (17.02 \times P_Q) + (11.78 \times P_C) + \\ &\quad (16.64 \times P_{SH}) + (11.73 \times P_F) + (7.74 \times P_{Co}) + (12.75 \times P_E) \qquad (6) \end{aligned}$$

DESCRIPTION OF THE FACTORS

Cost

Cost performance is result-oriented objective measure. It is defined as the completion of a project within the estimated budget. It can be measured by cost over-budget/under-budget. Percentage net variation over final cost gives an indication of cost over-budget or under-budget. Here assessment of cost by using five-point Likert scale (say within budget, on budget, or overrun budget). 0 to 5% reduction in actual construction cost than the planned cost represents "excellent performance" (Rating-5), 0% reduction in actual construction cost than the planned cost represents "good performance" (Rating-4), 0 to 5% increase in actual construction cost than the planned cost represents "average performance" (Rating-3), 5 to 10% increase in actual construction cost than the planned cost represents "poor performance" (Rating-2) and more than 10% increase in actual construction cost than the planned cost represents "extremely poor performance" (Rating-1).

Time

Time performance is defined as change in duration of construction with respect to the original planned duration. According to the client, user, stakeholder's completion of the project on time is their first criteria. Time performance can be measured in schedule variation. Assessment of time is done by using five point scale 0 to 5% early construction completion than the planned duration represents "excellent performance" (Rating-5), 0% delay in actual construction completion than the planned duration represents "good performance" (Rating-4), 0 to 5% increase in actual construction duration than the planned duration represents "average performance" (Rating-3), 5 to 10%

increase in actual construction completion time than the planned duration represents "poor performance" (Rating-2) and more than 10% increase in actual construction time than the planned duration represents "extremely poor performance" (Rating-1).

Client's Satisfaction

Client's satisfaction is by definition subjective, and as a consequence, is influenced by the individual client's satisfaction (The KPI Working Group, 2000). Clients' satisfaction is regarded as a function of comparison between an individual's perception of an outcome and its expectation for that outcome. In the construction industry client's satisfaction has remained an intangible and challenging issue. Dissatisfaction is widely experienced by clients of the construction sector and may be caused by many aspects but is largely attributable to overrunning project costs, delayed completion, inferior quality and incompetent service providers including contractors and consultants.

In the construction industry, the measurement of client's satisfaction is often associated with performance and quality assessment in the context of products or services received by the client. Usually, the client's requirements are to get construction-needs translated into a design that specifies characteristics, performance criteria and conformance to specifications, besides getting the facilities built within cost and time (Ali, 2010).

Client's satisfaction is measured subjectively using a five point scale, ranging from 1 = totally dissatisfied, 2 = somewhat dissatisfied, 3 = neutral (neither satisfied nor delighted), 4 = delighted, 5 = totally delighted.

Quality

Quality is defined as "the totality of features and characteristics of a product or service that bear on its ability to satisfy stated or implied needs". Quality is an important measure for projects which is often cited by researchers. Quality is subjective and means different things to different people. A quality issue may include incorrect information on drawing, defective materials, poor workmanship on site etc. At present there is no objective recognized method of measuring quality in the construction industry (The KPI Working Group, 2000).

Following scoring system was used for the measurement of quality; construction work which is apparently defect free and with good quality represents the "excellent quality" (Rating-5), construction work with few defects but not much effect on client represents "good quality" (Rating-4), construction work with some defects and some impact on client represents "average quality"

(Rating-3), major defects in construction work and much dissatisfaction of client represents "poor quality" (Rating-2) and if the construction work is totally defective and client is dissatisfied with the work represents "extremely poor quality" (Rating-2).

Safety and health

Safety is defined as the degree to which the general conditions promote the completion of a project without major accidents or injuries. It is a common practice that the measurement of safety performance mainly focused on the construction period because most accidents occur during this stage. Throughout the world, construction industry is known as one of the most hazardous industries. Thousands of people are killed and disabling injury annually in industrial accident. Construction workers worldwide have three times more chances of dying and two times of getting injured than any worker of other economic activity (Ali, 2010).

Following criteria are designed to measure safety: completed project with no accidents recorded then it is rated as "extremely safe" with the rating of 5, if some minor injuries were recorded then it is rated as "safe" with the rating of 4, project with major non-fatal injuries recorded is rated as "moderately unsafe" with the rating of 3, project with non-fatal accidents with permanent accident is recorded then rated as "unsafe" with the rating of 2, project with fatal accident recorded is rated as "extremely unsafe" with the rating of 1. If personal protective equipment is not used that will result reduction of one point in the score.

Communication

Communication is often adopted as a subjective measure for projects. Communication serves various functions, but its primary function is to convey information. The efficiency and effectiveness of the construction process strongly depend on the quality of communication.

However, related research suggests that communication is critical to the success of construction project teams. Project teams with better (e.g., frequent, detailed) communication experienced more positive outcomes (e.g., completed within budget, limited amount of rework). If this relationship can be generalized to construction communication, then better (e.g., relevant, accurate) construction communication will result in more positive outcomes (Robert, 2004).

If there were no complexity in communication and project was not affected due to communication then the communication performance rating is 5 (excellent performance), communication was good, there was some complexity but no delay and problems were recorded then the communication performance rating is 4 (good performance), some complexity in communication due to that some delay and problem were occurred in the project then the communication performance rating is 3 (average performance), much complexity in communication due to that many times delays and problems were occurred in the project then the communication performance rating is 2 (poor performance) and if the project was affected many time due to the not proper communication then performance rating is 1 (very poor).

Environment

Construction activity is generally considered as a major contributor to environmental pollution, as noted by various research works. Construction activity is one of the major contributors to the environmental impacts, which are typically classified as air pollution, noise pollution and water pollution. On-site construction activities relate to the construction of a physical facility, resulting in air pollution, water pollution, traffic problems and the generation of construction wastage.

For the assessment of the project environment three environmental factors were included in the research, which are air pollution, noise pollution and waste generation. The evaluation of environment performance is done by gathering information related to different criteria (listed below) and based upon these criteria evaluation of environment was carried out in five levels which are ranging from 1 = extremely poor performance, 2 = below average or poor performance, 3 = average performance, 2 = good performance to 1= excellent performance.

Table 6. List of questions for the rating of environment

Implementing a waste management plan.
Recycling/Reusing the construction and demolition waste.
Listing out the materials to be salvaged for reuse in the contract documents.
Coordinating the ordering and delivery of materials among all contractors and suppliers to ensure that the correct amount of materials is delivered and stored at the optimum time and place.
Providing access for waste collection vehicles to transfer the waste properly.
Providing facilities/places for sorting out wastes onsite.
Applying mitigation measures for controlling dust and air emissions.
Providing hoarding along the entire length of project site boundary.
Providing effective dust screen, sheeting, or netting to enclose scaffolding built around the perimeter of a building.

Using water sprays for watering unpaved areas, access roads, construction areas, and dusty stockpiles regularly to keep dusty surfaces wet.
Applying wire meshes, blast nets, or other covers on top of the blast area to prevent the flying off of rocks and to suppress dust generation during demolition.

Providing noise control measures.
Using mechanical and plumbing devices that generate less noise and dampen the noise generated during operation.
Avoiding locating mechanical equipment or noisy devices adjacent to noise-sensitive areas.
Preventing noise transmission by providing measures to absorb noise and vibrations.
Restricting nighttime working to low-noise activities to ensure no excess of acceptable noise level.
Operating noisy activity at times when adjacent area is more likely to remain unoccupied.

The criteria discussed for the factors above are applied to case studies and the project performance score is calculated using the performance index. Table 7 shows the summary of cases which includes rating of different factors and total project score.

Table 7. Summary of the cases

KPI	Weight	Case-1	Case-2	Case-3	Case-4	Case-5	Case-6	Case-7	Case-8	Case-9
Client's satisfaction	9.51	4	2	4	4	3	4	5	3	4
Time	12.83	1	1	3	1	3	1	5	2	4
Quality	17.02	5	3	3	4	3	5	5	4	5
Cost	11.78	3	1	3	2	2	2	4	2	4
Safety and health	16.64	4	3	3	4	2	4	5	4	4
Functionality	11.73	5	4	4	4	3	5	5	3	5
Communication	7.74	3	3	3	3	3	4	5	3	5
Environment	12.75	4	2	3	3	3	4	4	3	4
Score		**370.7**	**240.3**	**321.2**	**317.5**	**271.6**	**366.7**	**475.5**	**309.1**	**436.5**

Validation of Research Findings

The factors and their respective weights were calculated from the questionnaires but the validation of this research is needed for the verification of the index.

The score of the index was calculated on the project ranges from one crore to thirty crore. To determine the score that distinguishes between successful and unsuccessful projects several points (260, 270, ..., 310 and 320) were analysed using Mann-Whitney test by using SPSS software. According to the questionnaires and literature review, cost and time performance are the most important objective criteria which can be most helpful for the performance measurement. So, the values of time performance and cost performance were analysed at different points (260, 270, ..., 310 and 320) and it was seen that there is significant difference in the cost and time variation at the score of 320 when it was checked at 90% confidence level.

There was much difference in cost and time performance in the projects above scoring 320 and the projects scoring below 320. Table 8 and Table 9 shows the difference between various factors.

The performance index score for nine projects ranged from 240 to 475 with the mean score of 345. Four projects scored below 320 and five projects scored above 320.

Table 8. Analysis of cost and time variation for the validation

	Project score less than 320 (i)	Project score greater than 320 (ii)	Difference (iii) = (i) –(ii)
Mean rank of cost variation (%)	6.75	3.6	3.15
Mean rank time variation (%)	7.17	3.92	3.25
Average performance index score	284.5	394.2	-109.7

The above table shows the mean rank of cost variation and time variation of the projects scoring above and below 320; also average performance index score of the projects scoring above and below 320 is given. It is seen that cost and time variation of the projects scoring below 320 is 6.75 and 7.17 respectively, and cost and time variation of the projects scoring above 320 is 3.6 and 3.92 respectively. So the cost variation is 3.15% higher in projects scoring below 320 compared to the projects scoring above 320 and time variation is 3.25% higher in projects scoring below 320 compared to the projects scoring above 320. Also the mean of performance index score was analysed and it was found that there is difference of 109.7 when the cut-off of 320 was used for the analysis.

The other detail of the projects were also taken in the case studies which are quality performance, clients satisfaction, health & safety, environment, communication and functionality, they are also compared which is shown in Table 9.

Table 9. Analysis of different factors by using cut-off of 320

	Project scoring less than 320		Project scoring greater than 320		Difference between sum of scores	Difference between means
	sum	mean	sum	mean		
	(i)	(ii)	(iii)	(iv)	(v) = (iii) – (i)	(vi) = (iv) – (ii)
Cost performance score	7	1.75	16	3.2	9	1.45
Time performance score	7	1.75	14	2.8	7	1.05
Client satisfaction score	12	3	21	4.2	9	1.2
Quality performance score	14	3.5	23	4.6	9	1.1
Health & safety score	13	3.25	20	4	7	0.75
Environment score	11	2.75	19	3.8	8	1.05
Communication score	12	3	20	4	8	1
Functionality score	14	3.5	24	4.8	10	1.3
Project score	1138.3	284.5	1970.6	394.2	832.26	109.7

The above table shows the difference between the means of each performance criteria for the projects which scored above 320 and below 320. It is seen that all the factors (cost, time, client satisfaction, quality, health & safety, environment, communication, functionality) have performed well in the projects which have scored above 320. So the score 320 can be taken as a benchmark for the measurement of the projects.

Case-2, case-4, case-5 and case-8 has a project performance score below 320 and case-1, case-3, case-6, case-7 and case-9 has a project performance score above 320.

CONCLUSION

The first objective of this research is to identify the factors influencing project performance and from the analysis it is found that cost, time, functionality, quality, safety & health, communication, client's satisfaction and environment are the major factors which are important to measure the performance of the building projects.

The second objective is to generate the project performance index, from the above listed factors, performance index was generated and was applied to nine different projects ranging from one crore to thirty crore, after the validation of the project performance index it was seen that there is a difference in the scores of the factors when the project score were analysed at 320 score. So, it can be concluded that the project having score above 320 have good overall performance and project scoring below 320 have not performed well. Also the project scoring above 320 have good financial and time performance. There was also much difference in the scores of the other factors which are functionality, safety & health, communication, quality and environment.

Mann-Whitney test was used to analyse the result at 90% confidence. From this analysis it is concluded that project performance score of 320 can be set as a benchmark for the projects performance measurement of the building projects. The projects exceeding score of 320 can be considered as well performed or good projects and project does not exceeding the score of 320 can be considered as a project not performed well.

This project performance index is applied on the actual cases and also the validation of the performance index is done using SPSS. So, this performance index can be helpful to measure the performance of the building construction projects.

REFERENCES

Ali, A.S. and Rahmat, Ismail .(2010). "The Performance Measurement of Construction Projects Managed by ISO-Certified Contractors in Malaysia". *Journal of Retail & Leisure Property,* 9(1), pp. 25-35.

Done, R.S. (2004). "Improving Construction Communication", Arizona Department of Transportation, United States.

Hamilton, M.R. (1997). "Benchmarking Project Success". *Journal of Construction Education,* 2(1), pp. 66-76.

Khosravi, S. and Afshari, H. (2011). " A Success Measurement Model for Construction Projects". *International Conferences on Financial Management and Economics,* Singapore. 11, pp. 186-190.

Lam, E.W.M., Chan, A.P.C. and Chan, D.W.M. (2008). "Determinants of Successful Design-Build Projects". *J. Constr. Eng. Manage.,* 134(5), pp. 333-341.

Long, L.H., Jun, Y.L. and Young, D.L., (2008). "Delay and Cost Overruns in Vietnam Large Construction Projects: A Comparison with Other Selected Countries". *KSCE Journal of Civil Engineering,* 12(6), pp. 367-377.

Lotfi, F.H, Nematollahi, N., Behzadi, M.H. and Mirbolouki, M. (2010). "Ranking Decision Making Units with Stochastic Data by Using Coefficient of Variation". *Mathematical and Computational Applications,* 15(1), pp. 148-155.

Navon, R. (2005), "Automated Project Performance Control of Construction Projects". *Automation in Construction,* 14(4), pp. 467-476.

Rankin, J.H. and Willis, C.J. (2011). "Measuring the Performance of Guyana's Construction Industry using a Set of Project Performance Benchmarking Metrics". *Journal of Construction in Developing Countries,* 16(1), pp. 19-40.

Takim, R., Akintoye, A. and Kelly, J. (2003). "Performance Measurement Systems in Construction", *19th Annual ARCOM Conference*, 3-5 September 2003, University of Brighton. Association of Researchers in Construction Management,1, pp. 423-432.

The KPI Working Group (2000). "KPI Report for the Minister for Construction", Department of the Environment, Transport and the Regions.

Yang, H., Yeung, J.F.Y., Chan, A.P.C., Chiang, Y.H. and Chan, D.W.M. (2010) "A Critical Review Of Performance Measurement in Construction" . *Journal of Facilities Management*, 8(4), pp. 269-284.

3

Fuzzy Rule Base System Approach for Pipe Condition Assessment: Case of Surat City

Namrata Jariwala[1] and Hemali Tailor[2]

ABSTRACT

One of the greatest challenges facing municipal engineer is the condition assessment of buried infrastructure assets. It is mandatory process to establish management strategies for these assets. Condition assessment of water mains is challenging compared to other infrastructure assets because they are underground, operated under pressure and mostly they are inaccessible. The condition of water pipelines has been deteriorating with time and since this infrastructure is out of sight, the assessment has been neglected over the years. The advancement of technology in various fields has provided pathway for development of several technologies for assessment of the condition of pipeline systems. However, there is no standard guidance or tool for the utilities to use these technologies appropriately. The use of these approaches generates a quantitative picture of the condition and performance of main network towards the optimization of the maintenance and rehabilitation programmes.

In most of the Indian cities, there have been lots of leakages per day which increases the operation cost of water supply network. Risk of failure is defined as the combination of probability and impact severity of a particular circumstance that negatively impacts the ability of infrastructure assets to meet municipal objectives. The urban water supply is based on a large and complex infrastructure that has been expanded and developed during the last century. While getting older, water supply assets, primarily pipes, are exposed to the deterioration process, and consecutive pipe leakage and then failure of that pipe. It is common for cities to have scores, hundred and even more than a thousand water pipelines break each year. There is no measurement of water leakage from the pipeline in any Indian city because of lack of the information and data viability. In developing countries, the water network pipelines are in poor condition. The deterioration of pipes also affects operation and maintenance cost, water losses, frequent service disruption, and a reduction in the quality of water supplied.

To provide an adequate supply of safe water in a cost- effective, reliable and sustainable manner, it is essential that planner and authority must develop a

1. Assistant Professor, CED, SVNIT, Surat
2. Research Scholar, CED, SVNIT, Surat

clear understanding of water pipelines deterioration process. An accurate quantitative picture of the condition and performance of system will allow utilities to implement efficient proactive pipe leakage management strategies to minimize overall economic, social and environmental costs of water pipelines network operation.

Water utility operators manage and operate distribution systems in a reactive mode by responding to emergency break and water pipelines leakages. Experience has shown that a significant number of water line repairs are performed on an unscheduled basis. For planners, it is essential to predict accurately which components are in the most urgent need of repair, and when others will need to be repaired.

Looking to above situation, in this paper an attempt has been made to develop methodology for pipe condition assessment using Fuzzy Rule Based System Approach. The model is developed using MATLAB R2008a – fuzzy tool box. The present research assists in designing framework to evaluate the no. of LKGs in the existing pipe network. The system in present work considers the risk factors of leakages, mainly due to pipe age, pipe diameter, length of pipe, pipe thickness, operational pressure, type of traffic and depth of installation within three main categories physical factor, operational factor and environmental factor. A set of municipal water network data of South-West zone of Surat city are collected and used to examine the developed FRBS model. Total 493 sections of 62.5 km pipe network of Surat city has been analysed.

Keywords: *Water network, pipeline, no. of leakages, fuzzy rule base system (FRBS)*

INTRODUCTION

Based on the research problem and the literature review, deterioration of water pipelines and prediction of future failures are important issues in water network management and crucial factors in establishing the water pipelines renewal priorities. The aging of water supply infrastructure systems, coupled with the continuous stress placed on these systems by operational and environmental conditions, have led to their deterioration, which manifests itself in the following ways (Kleiner, 1997):

- Increased rate of pipe breakage due to deterioration in pipe structural integrity. This, in turn, causes increased operation and maintenance costs, increased loss of treated water, and social costs such as loss of service, disruption of traffic, disruption of business and industrial processes and disruption of residential life.

- Decreased hydraulic capacity of pipes in the systems, which results in increased energy consumption and disrupts the quality of service to the public.
- Deterioration of water quality in the distribution system due to the condition of inner surfaces of pipes which may result in taste, odour and aesthetic problems in the supply water and even public health problems in extreme cases.

It has been reported that the distribution system often involves 80% of the total expenditure in drinking water supply systems (Clark et al., 1988). Given the reality of scare capital resources, it is important that a comprehensive methodology be developed to assist planners and decision makers in finding the best rehabilitation policy that addresses the issues of safety, reliability, quality and efficiency.

METHODOLOGY FOR PRESENT RESEARCH

The following two step process highlights the selected methodology in this research:

1. Data collection
2. Development of model using MATLAB R2008a- fuzzy tool box

Study Area: Surat

The city of Surat is situated at latitude 21°12N and longitude 72°52E on the bank of river Tapi having coastline of Arabian Sea on its West. It is 13 m above the mean sea level. Surat city is located in well developed South Gujarat region. The city occupies a pivotal position on the Ahmedabad Mumbai regional corridor centrally located at a distance of 260 km. Here the study area of South-West zone of Surat city is selected for development of FRBS model for pipe condition assessment. Area covering the South-West zone is 111.912 sq. km. and population of the South-West zone is 2.42 lacs. Figure 1 shows the map of the South-West zone of Surat city.

The area chosen for this study, South-West zone of Surat city (India) is an extraordinary case where the pipelines age varies from 5 to 31 years. In South-West zone, T.P.-27, 28 and 6 water distribution network data are collected and used for FRBS model development. The draft TP scheme map is as in Figure 2 and 3.

The SMC operates a water distribution system providing retail water service to more than 2,42,466 (as per 2001) people (South-West zone). The majority of pipe material in this zone is cast iron, mild steel and ductile iron.

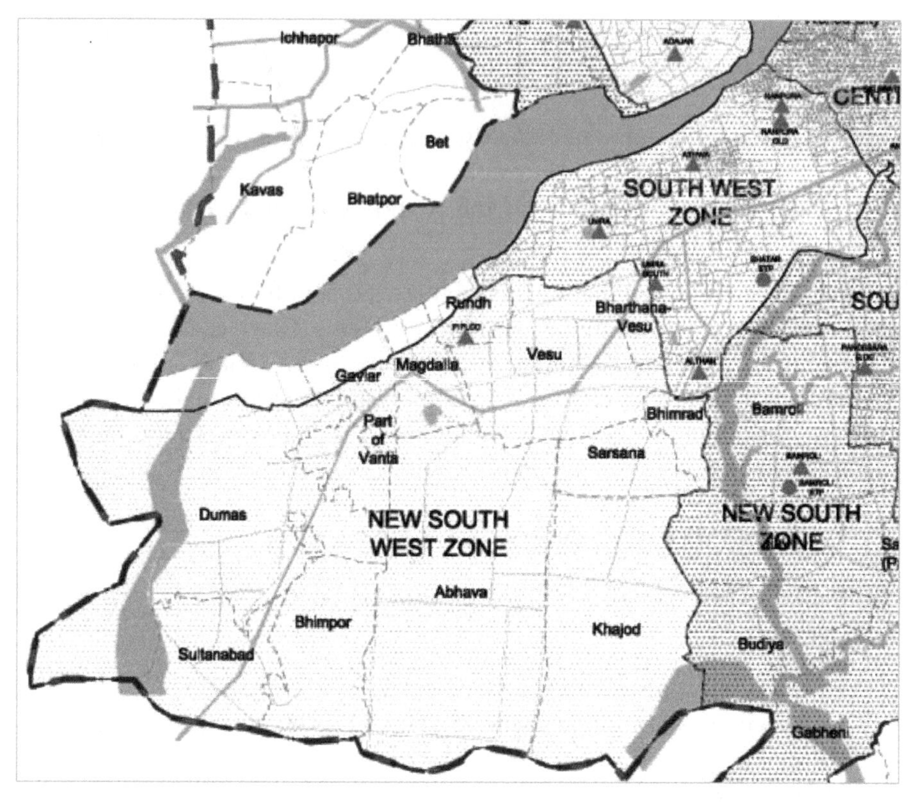

Source: SMC, Surat

Fig. 1. Map showing the South-West zone of Surat city

Recently, polyethylene pipes have been used extensively for new connection. Average water supply for this zone is approximately 760 lac litre daily (as per 2010). Treated water is delivered from the water treatment plant to the elevated reservoirs. The pressure in the network is found between 0.5 to 3.5 kg/cm^2. The pipe line network map is as given in Figure 4.

Data Collection for South-West zone

From the year in which the installation of water network took place, the age of the pipes can be considered. Age of the pipe data was collected from zonal hydraulic office. Diameter and length of the pipes were directly taken from the water distribution network map of particular T.P. scheme. From IS-code: 8329: 2000 the thickness of the pipe data is available based on the diameter of the pipe. Operational pressure to be maintained at the different section of the pipes network is available from the leakage inspector. The type of traffic at a

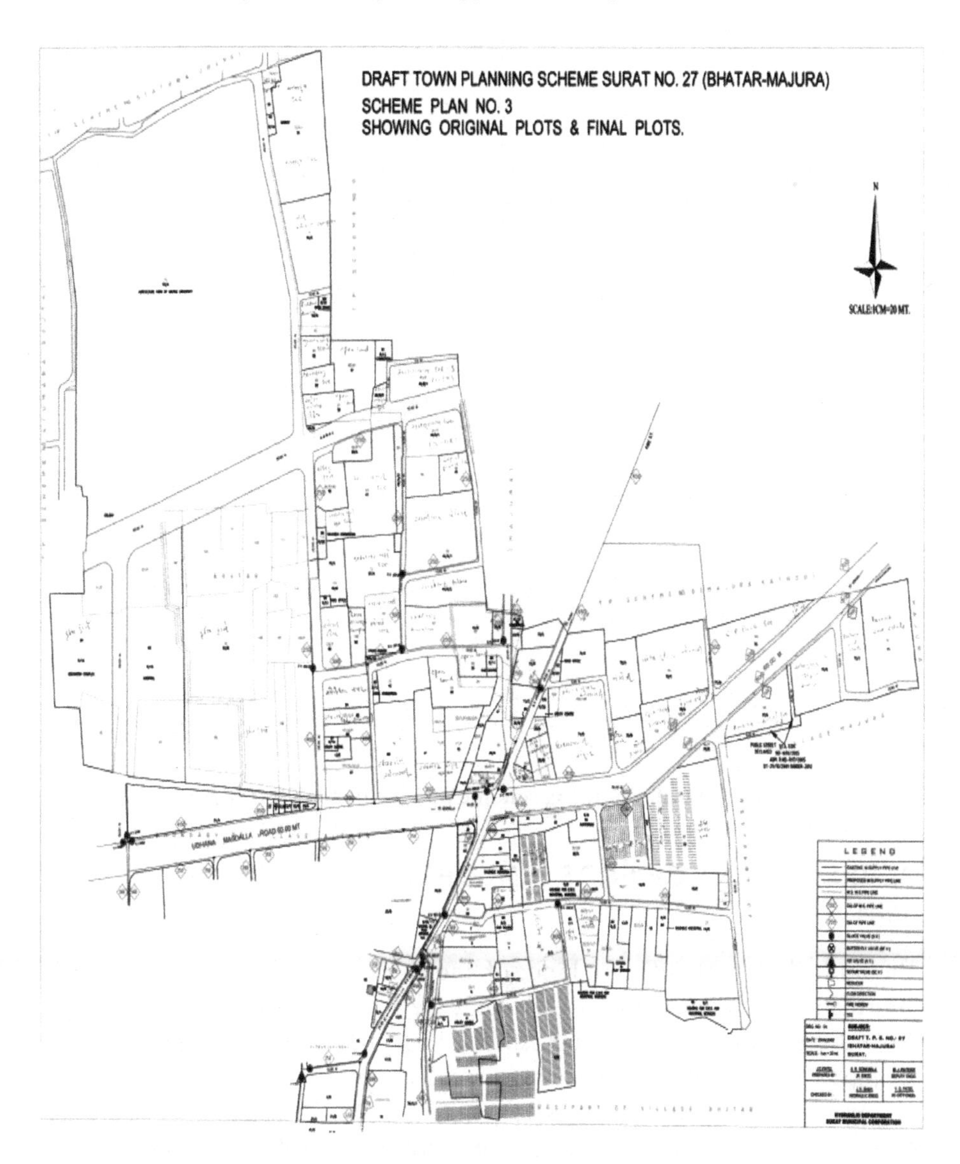

Fig. 2. Draft planning scheme of TP No. 27 of Surat city

particular section is classified as low, medium and high traffic based on the type of road and capacity of the traffic (PCU). Depth of installation is based on the traffic capacity and collected from the leakage inspector of South-West zone of Surat city.

Total 493 pipe section data were collected and used for development of FRBS model for pipe condition assessment.

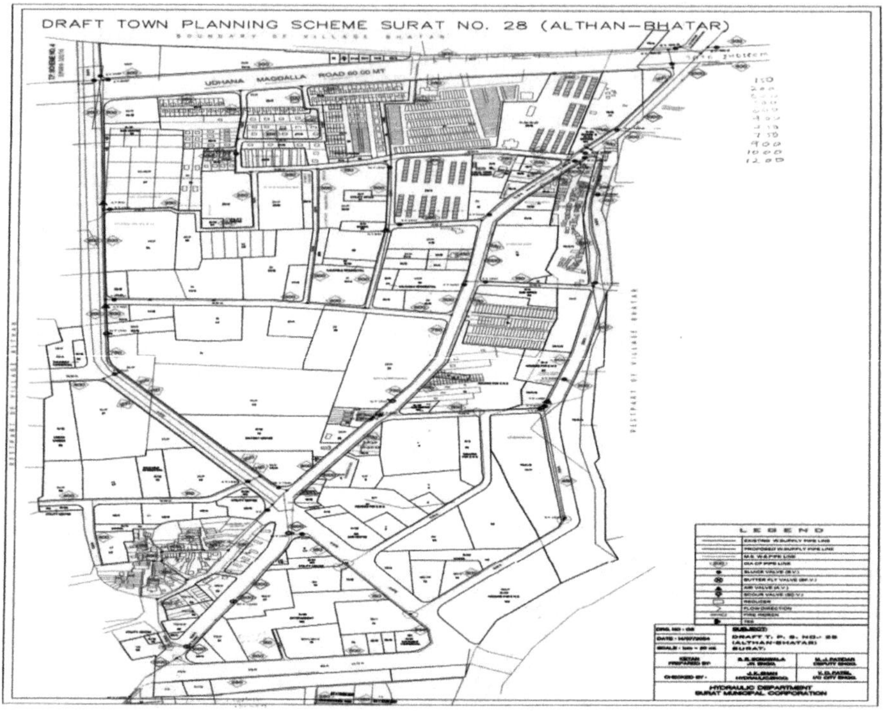

Fig. 3. Draft planning scheme of TP No. 28 of Surat city

DEVELOPMENT OF FRBS FOR DETERMINATION OF PIPE CONDITION

The study deals with different kinds of knowledge acquisition way to establish information of water network that could be beneficiary in developing fuzzy linguistic parameters and their associated membership function to quantify the pipe condition of occurrences of no. of leakages.

As objective of the study is to get No. of LKGs. in water distribution network, first the important factors responsible for leakages in the network must be identified. This can be analysed from available literature review and discussion with the experts dealing with the water distribution network.

The composite structure for the assessment of water pipe condition is considered as presented in the Figure 5.

Defining the Fuzzy Input and Output Parameter

The linguistic variable for each input and output parameters and their range is selected with the help of the experts. The questionnaire is prepared and the

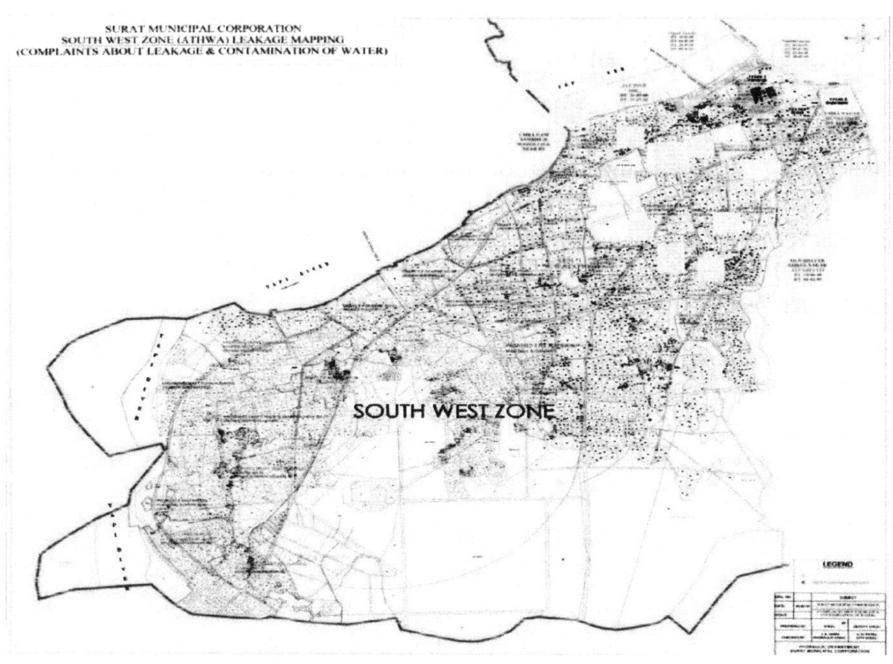

Fig. 4. South west zone water pipelines network

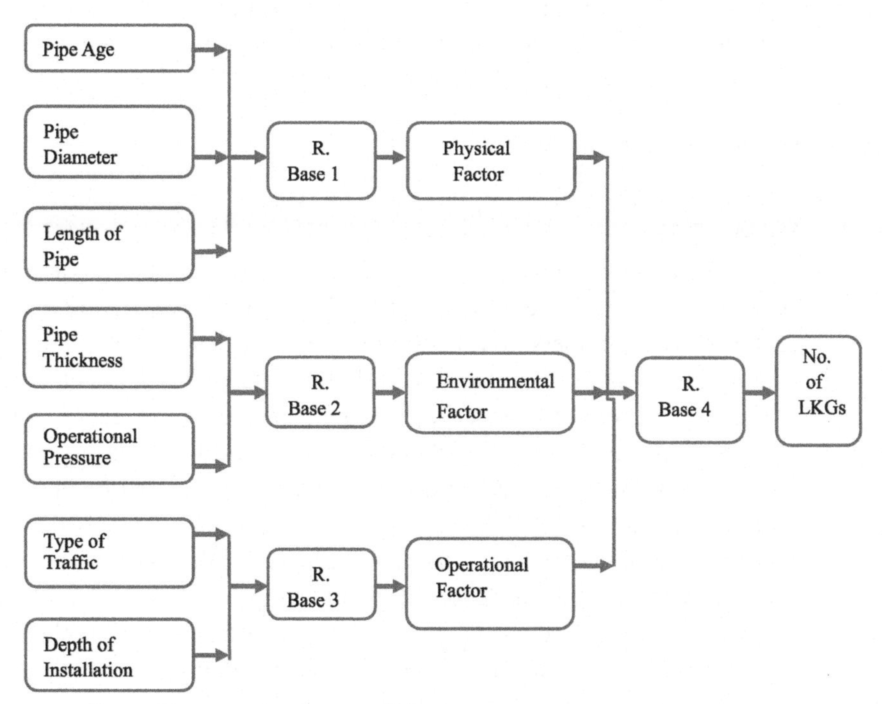

Fig. 5. Water main pipe condition assessment composite structures

effective range of variables is asked to the experts. The input parameters of fuzzy systems are decided from the raw data and judgement of experts. The pipe age, diameter and length of pipe is the input parameter for physical factor. Pipe thickness and operational pressure are input parameter for the operational factor. Type of traffic and depth of installation is the input parameter for the environmental factor. Pipe material and C-factor is almost constant for all the data, so two input parameters are omitted in the FRBS model.

Characterization of the Fuzzy Input Parameters

Subjective judgement, gathered data and the linguistic variables are employed to develop fuzzy membership functions of all input parameters and output parameters.

An element can only be inside or outside the set while element in fuzzy logic can be partially or totally inside or outside. The position of the element is described by the membership value (μ) that has a value of one (μ=1), if the element belongs completely to the set and a value of zero (μ=0) if the element does not belong to the fuzzy set and any value between zero and one ($0<\mu<1$) when the element belongs, partially to the fuzzy set. Additionally we can define the degree of membership in linguistic fuzzy set for any precise value. A membership function assigns a value between 0 and 1 for each. The most convenient and simplest membership functions are formed using straight lines like triangle and trapezoidal membership functions.

Determination of Physical Factor

- Three inputs are defined for physical factor: pipe age, pipe diameter and length of pipe.
- Valid ranges of the inputs are considered and divided into classes, or fuzzy sets. For example, pipe age can range from "New" "Moderate" and "old". The membership function for the pipe age is shown in Figure 6 and their ranges are presented in Table 1. The pipe diameter and length of pipe both can range from "small", "medium" and "large". The membership function for the diameter of pipe and the length of pipe are shown in Figure 7 and 8 and their ranges are presented in Table 2 and Table 3 respectively. We can not specify clear boundaries between classes, the degree of belongingness of the values of the variables to any selected class is called the degree of membership;
- The output is physical factor and is defined in fuzzy sets like "very risky", "risky", "adequate", "good" and "excellent". Membership function

for the physical factor is shown in Figure 9 and the linguistic variables for physical factors are selected by considering literature review and expert opinion as shown in Table 4.

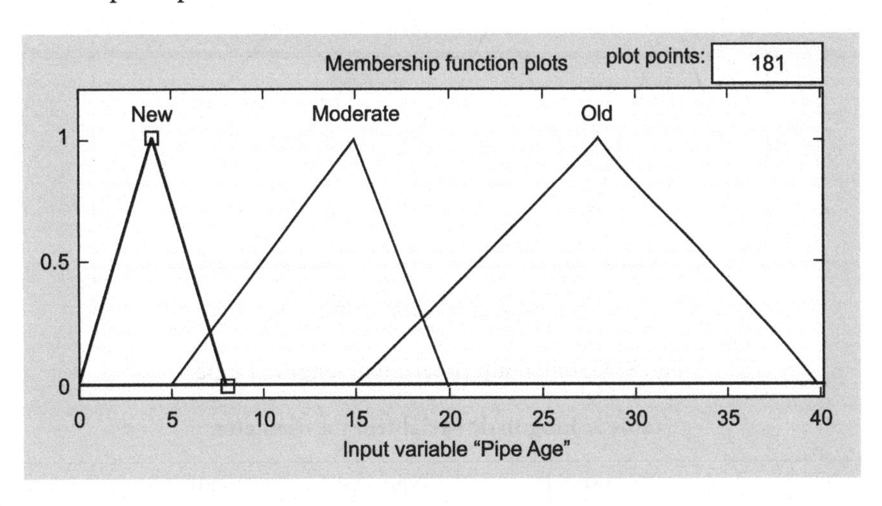

Fig. 6. Membership function of pipe age

Table 1. Linguistic variable: Pipe age

Sr. No.	Linguistic variable	Type of M.F.	Assigning fuzzy number
1.	New	Triangle	(0, 4, 8)
2.	Moderate	Triangle	(5, 15, 20)
3.	Old	Triangle	(15, 28, 40)

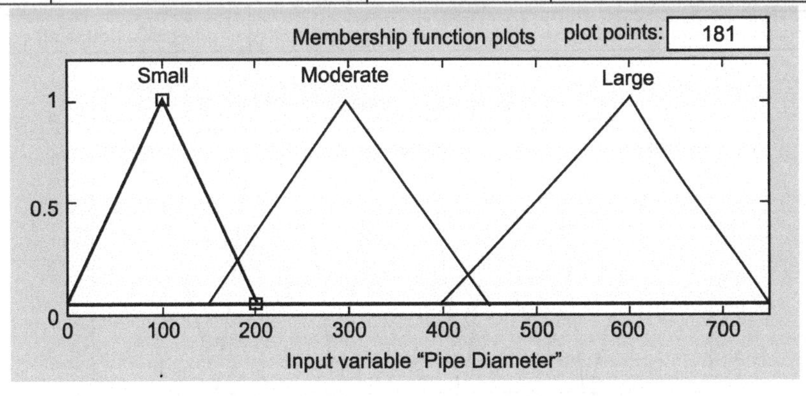

Fig. 7. Membership function of pipe diameter

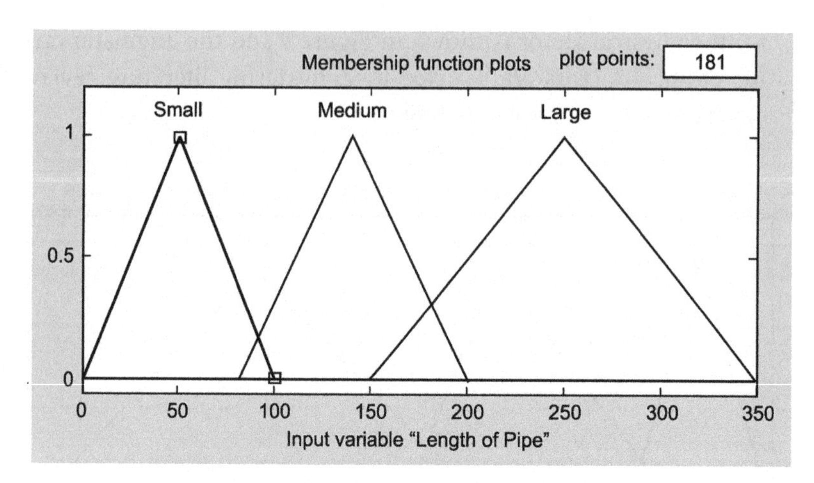

Fig. 8. Membership function of length of pipe

Table 2. Linguistic variable: Pipe diameter

Sr. No.	Linguistic variable	Type of M.F.	Assigning fuzzy number
1.	Small	Triangle	(0, 100, 200)
2.	Medium	Triangle	(150, 300, 450)
3.	Large	Triangle	(400, 600, 750)

Table 3. Linguistic variable: Length of pipe

Sr. No.	Linguistic variable	Type of M.F.	Assigning fuzzy number
1.	Small	Triangle	(0, 50, 100)
2.	Medium	Triangle	(80, 140, 200)
3.	Large	Triangle	(150, 250, 350)

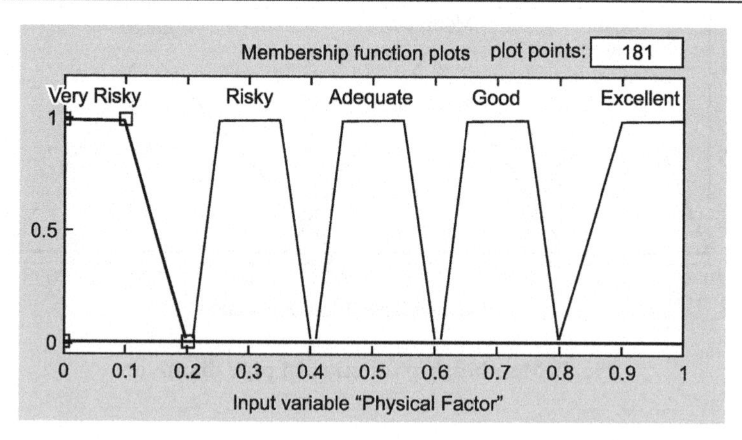

Fig. 9. Membership function for output of physical factor

Table 4. Linguistic variable: Physical factor

Sr. No.	Linguistic variable	Type of M.F.	Assigning fuzzy number
1.	Very Risky	Trapezoidal	(0, 0, 0.1, 0.2)
2.	Risky	Trapezoidal	(0.21, 0.25, 0.35, 0.4)
3.	Adequate	Trapezoidal	(0.41, 0.45, 0.55, 0.6)
4.	Good	Trapezoidal	(0.61, 0.65, 0.75, 0.8)
5.	Excellent	Trapezoidal	(0.8, 0.9, 1, 1)

Determination of Operational Factor

- Two inputs are defined for operational factor: pipe thickness and length of pipe.
- Valid ranges of the inputs are considered and divided into classes, or fuzzy sets. For example, pipe thickness can range from "thin", "moderate" and "thick". The membership function for the pipe thickness is shown in Figure 10 and their ranges are presented in Table 5. The operational pressure can range from "High pressure",, "Medium pressure", and "Low pressure". The membership function for the operational pressure is shown in Figure 11 and their ranges are presented in Table 6.
- The output is operational factor and is defined in fuzzy sets like "very risky", "risky", "moderate", "good" and "excellent". The membership function for the operational factor is shown in Figure 12 and the linguistic variables and their ranges of operational factor are selected by the literature review and with the help of the experts, which is same as in table 4 and given in Table 7.

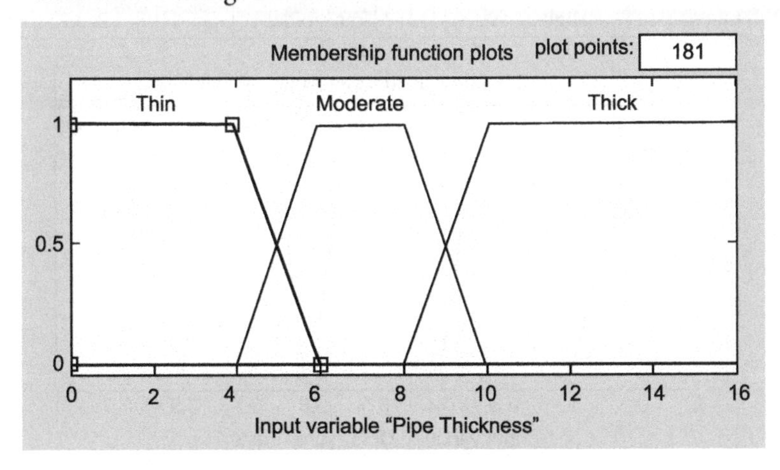

Fig. 10. Membership function of pipe thickness

Table 5. Linguistic variable: Pipe thickness

Sr. No.	Linguistic variable	Type of M.F.	Assigning fuzzy number
1.	Thin	Trapezoidal	(0, 0, 4, 6)
2.	Moderate	Trapezoidal	(4, 6, 8, 10)
3.	Thick	Trapezoidal	(8, 10, 16, 16)

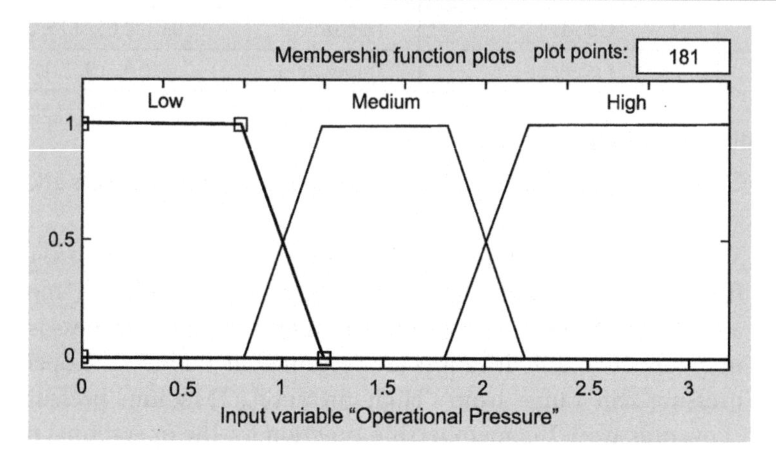

Fig. 11. Membership function of operational pressure

Table 6. Linguistic variable: Operational Pressure

Sr. No.	Linguistic variable	Type of M.F.	Assigning fuzzy number
1.	Low	Trapezoidal	(0, 0, 0.8, 1.2)
2.	Medium	Trapezoidal	(0.8, 1.2, 1.8, 2.2)
3.	High	Trapezoidal	(1.8, 2.2, 3.2, 3.2)

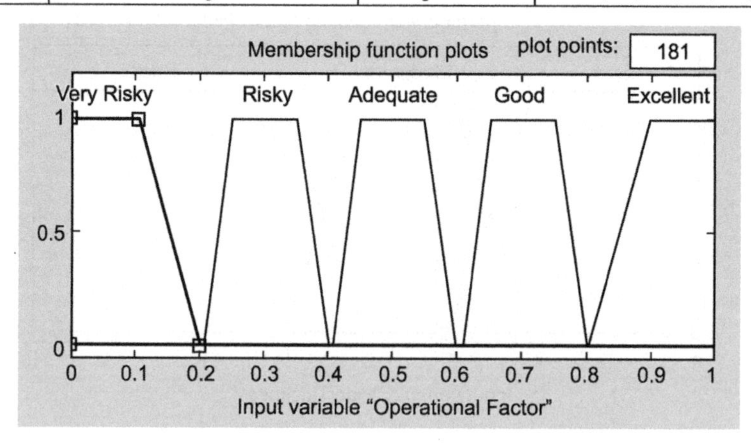

Fig. 12. Membership function for output operational factor

Table 7. Linguistic variable: Operational factor

Sr. No.	Linguistic variable	Type of M. F.	Assigning fuzzy number
1.	Very Risky	Trapezoidal	(0, 0, 0.1, 0.2)
2.	Risky	Trapezoidal	(0.21, 0.25, 0.35, 0.4)
3.	Adequate	Trapezoidal	(0.41, 0.45, 0.55, 0.6)
4.	Good	Trapezoidal	(0.61, 0.65, 0.75, 0.8)
5.	Excellent	Trapezoidal	(0.8, 0.9, 1, 1)

Determination of Environmental Factor

- Two inputs are defined for environmental factor: type of traffic and depth of installation
- Valid ranges of the inputs are considered and divided into classes, or fuzzy sets. For example, type of traffic can range from "low", "moderate" and "heavy". The membership function for the type of traffic is shown in Figure 13 and their ranges of the linguistic variables are presented in Table 8. The depth of installation can ranges from "low", "medium" and "high". The membership function for the depth of installation is shown in Figure 14 and their ranges of the linguistic variables are presented in Table 9.
- The output is environmental factor and is defined in fuzzy sets like "very risky", "risky", "moderate", "good" and "excellent". The membership function for the environmental factor is shown in Figure 15 and linguistic variable and ranges of environmental factor are selected by the literature review and experts as presented and mentioned in Table 10.

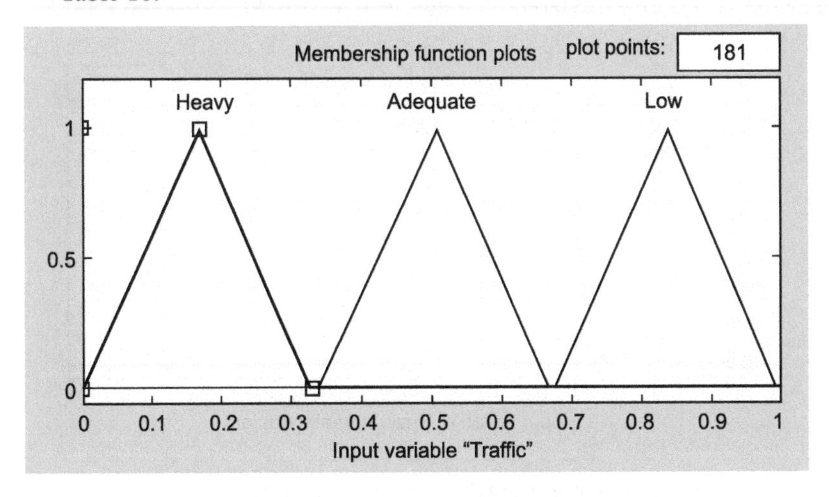

Fig. 13. Membership function for type of traffic

Table 8. Linguistic variable: Type of traffic

Sr. No.	Linguistic variable	Type of M.F.	Assigning fuzzy number
1.	Heavy	Triangle	(0, 0.17, 0.33)
2.	Moderate	Triangle	(0.34, 0.51, 0.67)
3.	Low	Triangle	(0.68, 0.84, 1)

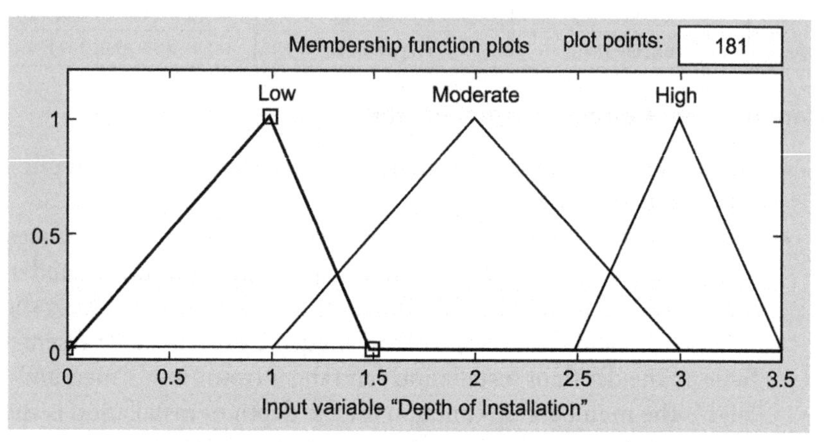

Fig. 14. Membership function of depth of installation

Table 9. Linguistic variable: Depth of installation

Sr. No.	Linguistic variable	Type of M.F.	Assigning fuzzy number
1.	Low	Triangle	(0, 1, 1.5)
2.	Moderate	Triangle	(1, 2, 3)
3.	High	Triangle	(2.5, 3, 3.5)

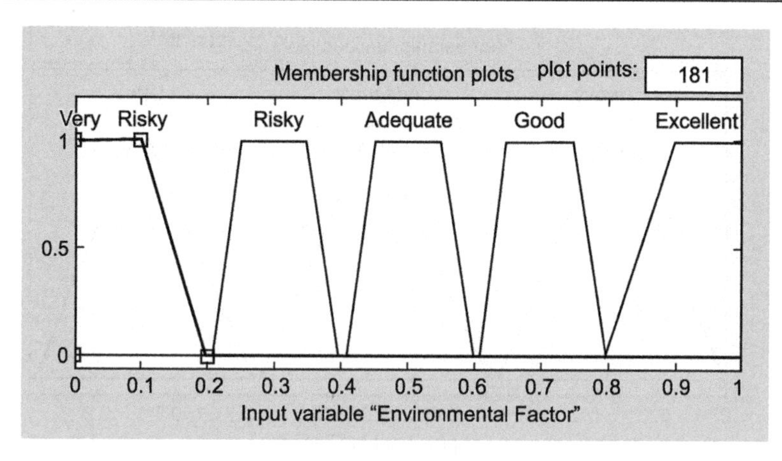

Fig. 15. Membership function for output environmental factor

Table 10. Linguistic variable: Environmental factor

Sr. No.	Linguistic variable	Type of M.F.	Assigning fuzzy number
1.	Very Risky	Trapezoidal	(0, 0, 0.1, 0.2)
2.	Risky	Trapezoidal	(0.21, 0.25, 0.35, 0.4)
3.	Adequate	Trapezoidal	(0.41, 0.45, 0.55, 0.6)
4.	Good	Trapezoidal	(0.61, 0.65, 0.75, 0.8)
5.	Excellent	Trapezoidal	(0.8, 0.9, 1, 1)

Determination of Final Pipe Condition (No. of LKGs.)

- Three inputs are defined for pipe condition: physical factor, operational factor and environmental factor.
- Valid ranges of the inputs of the no. of LKGs. are considered and divided into classes, or fuzzy sets. For example, physical factor, operational factor and environmental factor of the pipe conditions are define in fuzzy sets like "very risky", "risky", "adequate", "good" and "excellent". The membership function for the all three factors are shown in Figures 9, 12, and 15 respectively and their ranges of the linguistic variables are presented in Table 4 , Table 7 and Table 10 respectively.
- The output is no. of LKGs. and is define in fuzzy sets like "excellent", "good", "moderate", "risky" and "very risky". The membership function for the no. of LKGs. is shown in Figure 16 and their ranges of the linguistic variables are presented in Table 11.
- Expert knowledge is used to characterize inputs and outputs and connect the inputs and outputs by a set of inference rules using if-then statements; and
- The fuzzy output set is then defuzzified to arrive at a scalar value. The linguistic variable and ranges of pipe condition (No. of LKGs.) are selected by the literature review and experts as shown in different tables.

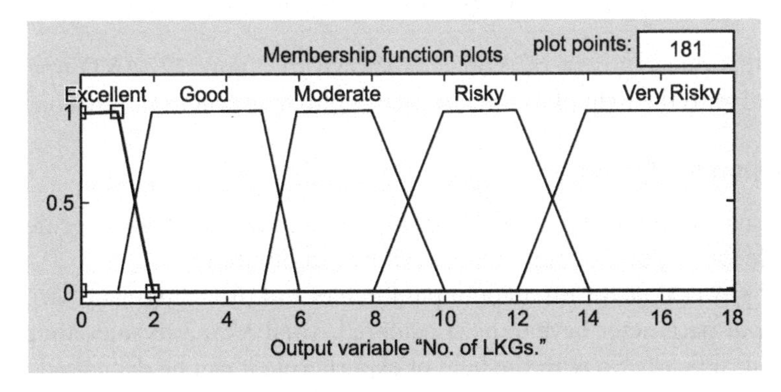

Fig. 16. Membership function for output no. of leakages

Table 11. Linguistic variable: Pipe condition (No. of LKGs.)

Sr. No.	Linguistic variable	Type of M.F.	Assigning fuzzy number
1.	Excellent	Trapezoidal	(0, 0, 1, 2)
2.	Good	Trapezoidal	(1, 2, 5, 6)
3.	Adequate	Trapezoidal	(5, 6, 8, 10)
4.	Risky	Trapezoidal	(8, 10, 12, 14)
5.	Very Risky	Trapezoidal	(12, 14, 18, 18)

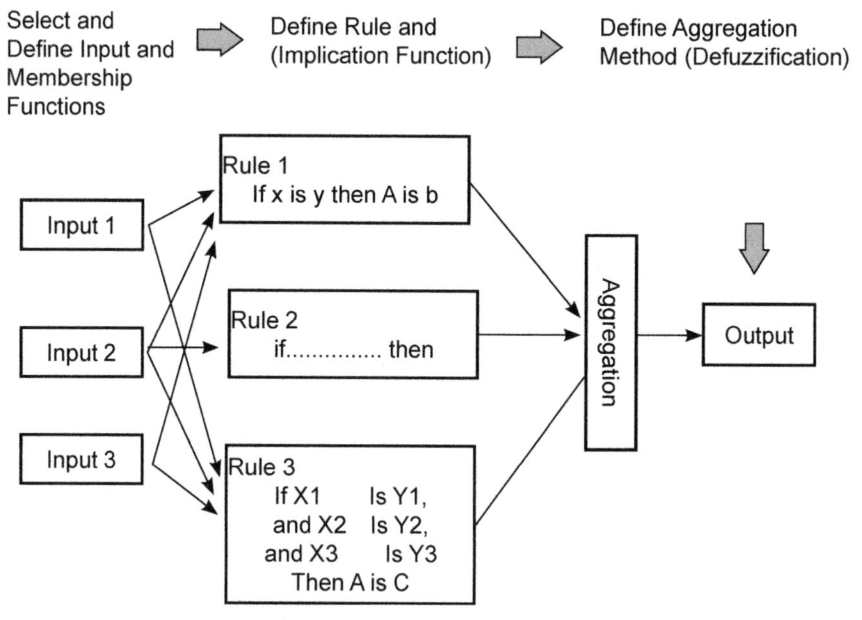

Fig. 17. Building a FRBS model

In this development of model, as shown in Figure 17. AND operator is used and with the help of Mamdani method further evalutions are considered.

Development of Rules

As we have developed the model with MATLAB–fuzzy tool box the rule viewer available for physical factor will be as shown in Figure 18.

For physical factor three input parameters and three linguistic variables of each input parameter have to be considered. Total 3×3×3×5 rules are possible for physical factor. So with the help of experts rule it can be decreased upto 27 which is represented in Figure 18.

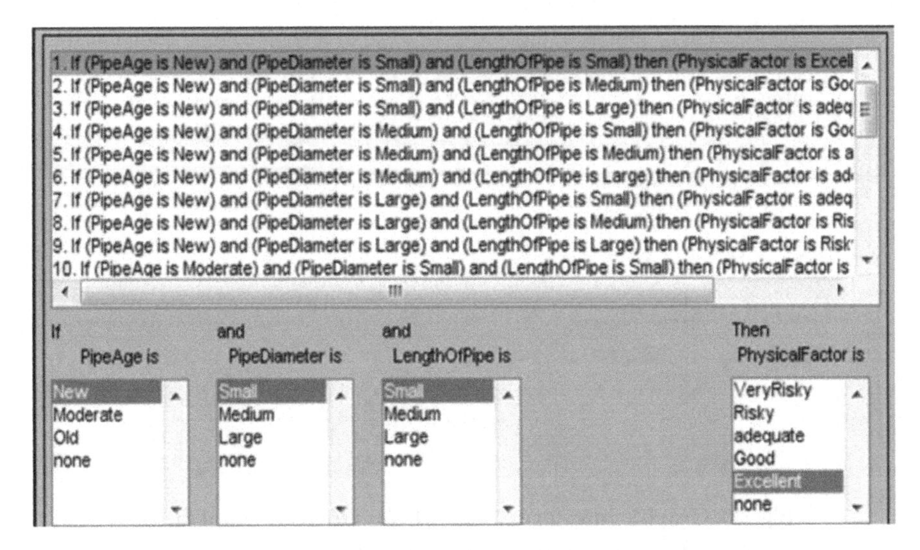

Fig. 18. Graphic interface of the rule for physical factor

For operational factor two input parameters and three linguistic variables of each input parameter have to be considered. Total 3×3×5 rules are possible for operational factor. So with the help of experts rule it can be decreased upto 9 which is represented in Figure 19.

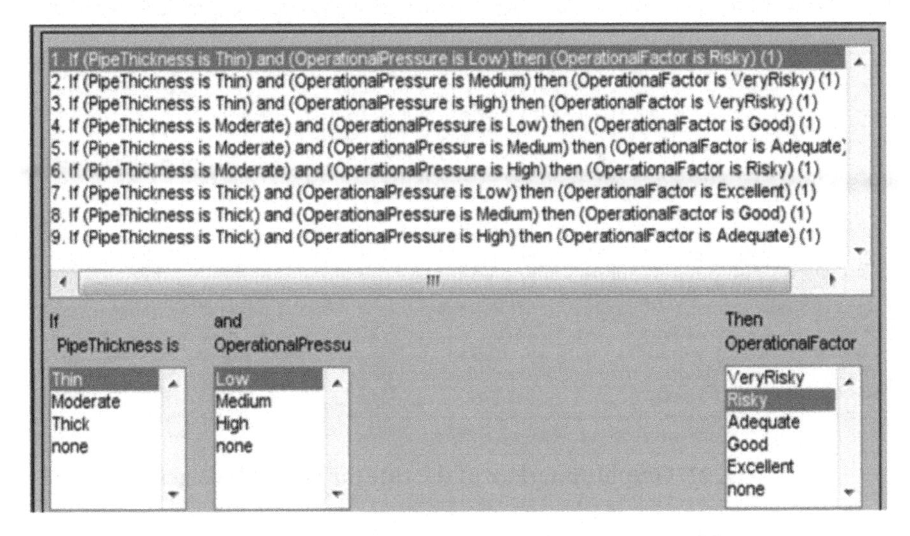

Fig. 19. Graphic interface of the rule for operational factor

For environmental factor two input parameters and three linguistic variables of each input parameter have to be considered. Total 3×3×5 rules are possible for environmental factor which is represented in Figure 20.

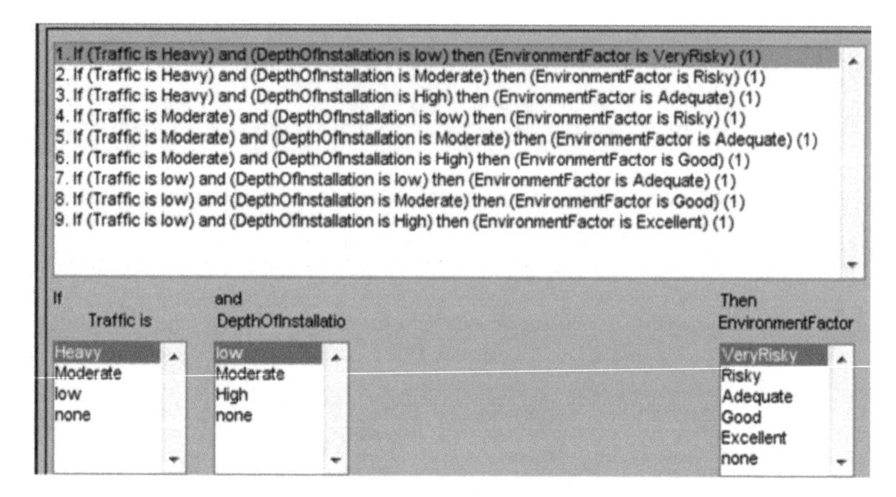

Fig. 20. Graphic interface of the rule for environmental factor

For no. of LKGs. three input parameters and five linguistic variables of each input parameter have to be considered. Total 5×5×5×5 rules are possible for no. of LKGs. So with the help of experts rule it can be decreased upto 125 which is represented in Figure 21.

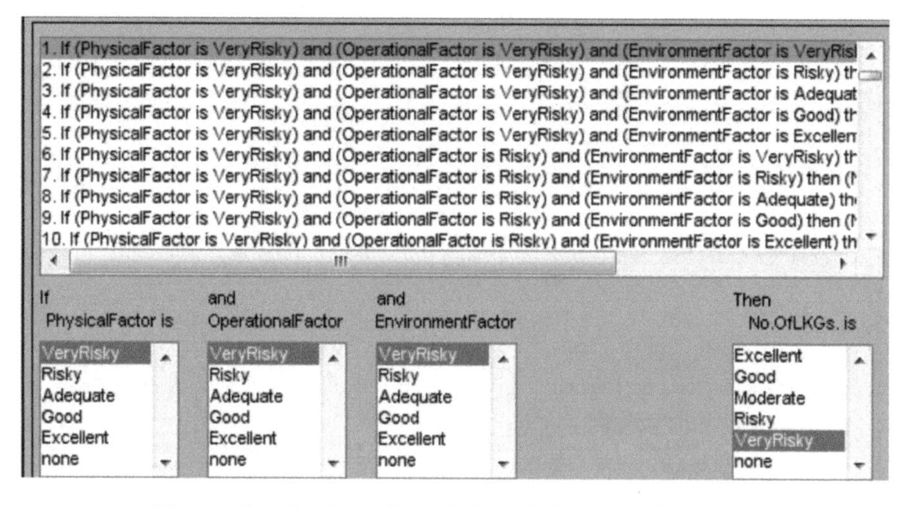

Fig. 21. Graphic interface of the rule for nos. of leakages

Defuzzification

The most popular defuzzification method is the centroid method used in MATLAB tool box and output of the system is obtained.

If we put the data of pipe age, diameter and the length of pipe from that we can get the physical factor by using Mamdani inference method in FRBS model with the help of the software MATLAB R2008a–fuzzy tool box. The simulation of the rule base and Mamdani method for inference of the physical factor are presented in Figure 22.

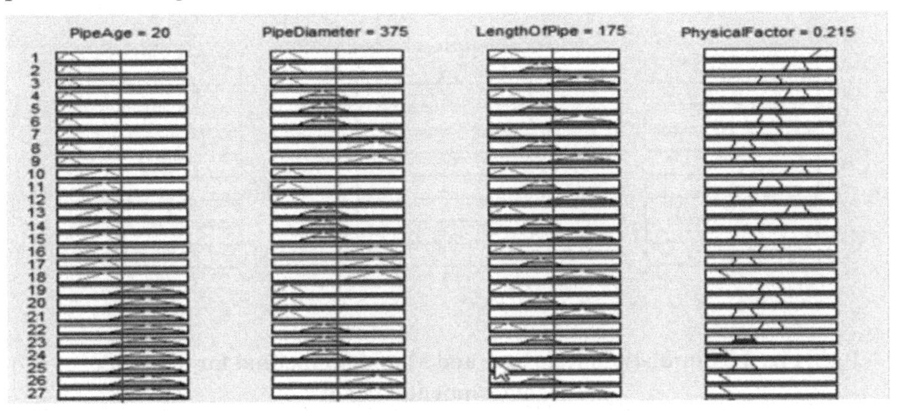

Fig. 22. Simulation: Rule base and Mamdani method for inference of physical factor

If we put the data of pipe thickness and operational pressure from that we can get the operational factor by using Mamdani inference method in FRBS model with the help of the software MATLAB R2008a–fuzzy tool box. The simulation of the rule base and Mamdani method for inference of the operational factor are presented in Figure 23.

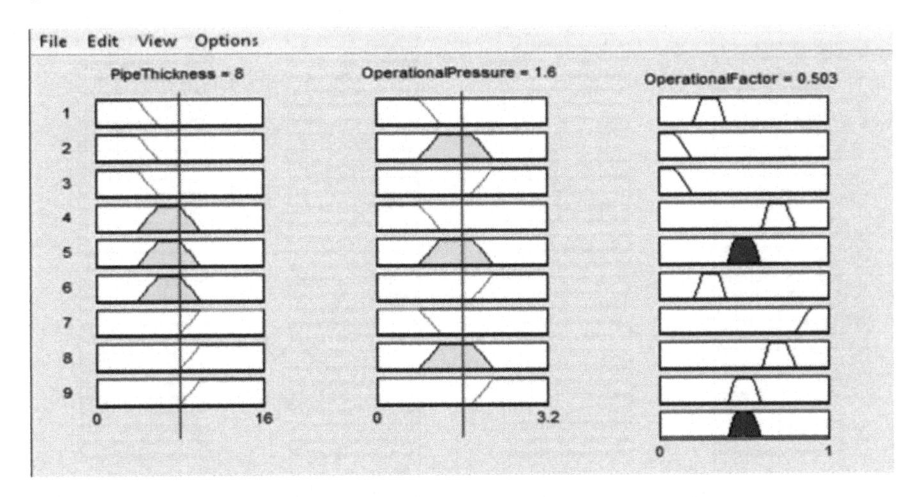

Fig. 23. Simulation: Rule base and Mamdani method for inference of operational factor

If we put the data of type of traffic and the depth of installation, from that we can get the environmental factor by using Mamdani inference method in FRBS model with the help of the software MATLAB R2008a–fuzzy tool box. The simulation of the rule base and Mamdani method for inference of the environmental factor are presented in Figure 24.

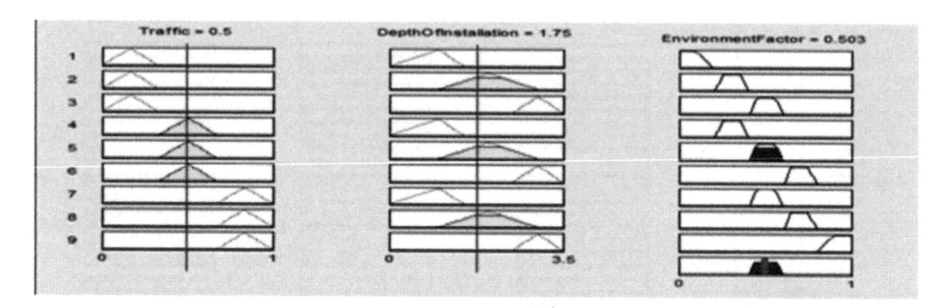

Fig. 24. Simulation: Rule base and Mamdani method for inference of environmental factor

If we put the data of physical, operational and environmental factor, from that we can get the no. of LKGs. by using Mamdani inference method in FRBS model with the help of the software MATLAB R2008a–fuzzy tool box. The simulation of the rule base and Mamdani method for inference of the No. of LKGs. are presented in Figure 25. For this particular work four different sets of fuzzy rules are defined.

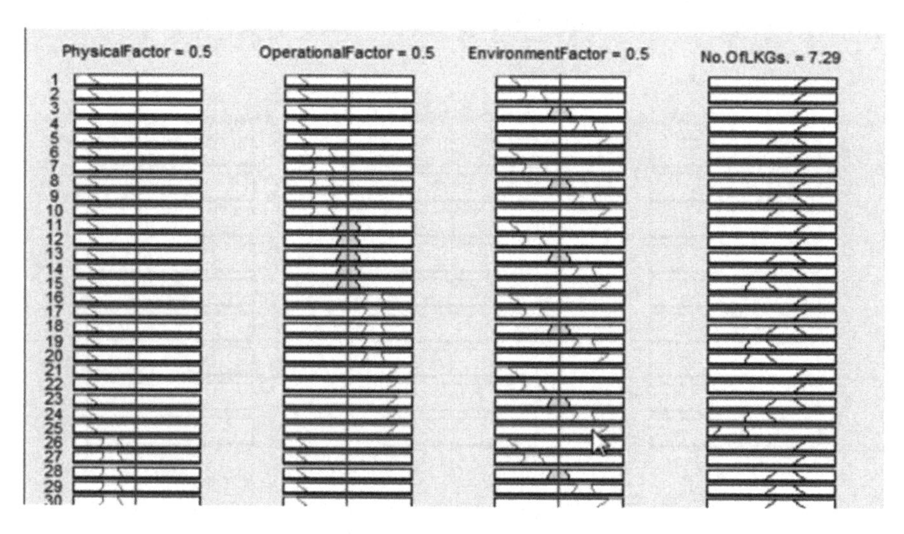

Fig. 25. Simulation: Rule base and Mamdani method for inference of nos. of leakages

Hence, we used Mamdani inference in which fuzzy sets from the consequent of each rule are combined through the aggregation operator and the resulting fuzzy set is defuzzified to yield the output of the system. Many practical approaches can be used together. Information and knowledge required in driving fuzzy rules and most commonly used techniques are statically data and information analysis, expert experience and engineering knowledge analysis, concept mapping and fuzzy modelling.

All the rules that have any truth in their premises will contribute to the fuzzy risk level that expression. Each single rule is fired to a degree that is a function of the degree to which the antecedent matches the input. This imprecise matching leads to interpolation between possible input states and serves to minimize the number of rules for describing input output relation.

The input parameter of the no. of leakage is physical, operational and environmental factors. The physical, operational and environmental factors are defined as having a range of [0, 1]. Here 0 represents the most risky condition of the pipeline and 1 represents good condition of pipeline. Here, physical, operational and environmental factors are inversely proportional to the no. of leakages of the pipe.

Output of the system, the no. of leakages is defined as having a range of [0, 18]. Zero(0) represents the excellent condition of pipe where the 18 represents the most risky condition of the pipes and they attempt immediate action for the rehabilitation of the pipes.

A simplified model is presented in this research having four steps: first step having 3 input and 1 output with 27 inference rules, second step having 2 input and 1 output with 9 inference rules and third step having 2 input and 1 output with 9 inference rules, and finally fourth step having 3 input and 1 output with 125 inference rules. In reality, there are no limits on the number of inputs, or outputs or the number of rules or the number of classes used to define the range of a variable. The model can be refined by adding more inputs and more rules. The modelling process is made convenient by fuzzy logic toolbox (Version R2008a).

The present study is designed to determine the no. of LKGs. in South-West zone of Surat city. Data of T.P. scheme 27 and 28 is used and from the developed model the no. of LKGs. in water distribution network is predicted.

In present research work total 493 important sections of 62.5 km length water distribution network of South-West zone particularly T.P.-27 and 28 is analysed and used for development of FRBS model and leakages is predicted in this section.

From the analysis depending upon the condition of physical, operational and environmental factors the leakages is obtained in the range of 4–16. The analysed water distribution network can be summarized in Table 12.

Table 12. Pipe condition of T.P.-27, 28

Pipe section	Excellent	Good	Moderate	Risky	Very Risky
493	0 (0%)	4 (0.81%)	180 (36.51%)	266 (53.96%)	43 (8.72%)

Physical factor, operational factor and environmental factor have a predominant role for pipe condition assessment.

CONCLUSION

The present study is designed to determine the no. of LKGs. of the water main pipes of South-West zone of Surat city. The no. of leakages of the water main pipes for the present research found out from 3.5 to 15.5, further it can be categorized as excellent, good, moderate, risky and very risky.

The present study showed that, out of 493 sections of 62.5 km of pipes are analysed, 0.81% (0.28 km) of pipes are in good condition. Likewise 36.51% (22.695 km), 53.96% (32.64 km) and 8.72% (6.885 km) of pipes are in moderate, risky and very risky condition respectively.

The present research summarized that most of the pipes are in moderate and risky condition. So there is need for replacement of the pipes is require an immediate action.

From the results we can see that newly installed pipes having minimum no. of LKGs. are in good condition. So the pipe age is the dominating factor for pipe condition assessment. In some of the data the pipe age is not dominating factor for assessment of the pipe condition.

If the municipalities or urban authorities are maintaining the data for the water network and manage the data of operating condition of water mains, prediction of the no. of leakages in water main pipeline is possible, using FRBS approach. From that they can easily maintain the water network and save our natural resources.

REFERENCES

Al-Barqawi, H. and Zayed, T. (2008). "Infrastructure Management: Integrated AHP/ANN Model to Evaluate Municipal Water Mains' Performance". *Journal of Infrastructure Systems*, 14(4), pp. 305-318.

Asnaashari, Ahmad (2007). "Water Pipelines Failure Modeling: Statistical, Artificial Neural Networks and Survival Modeling". Ph.D. report.

Clark, R.M., Eilers, R.G., and Goodrich, J.A. (1988). "Distribution System: Cost of Repair and Replacement". *Proc.,Conf. on Pipeline Infrastructure*, B.A. Bennett, ed., ASCE, New York, 428-440.

Dombi, J. (1990). "Membership Functions as an Evaluation". *Fuzzy Sets and Systems*, 35, pp.1-21.

Fares, H. and Zayed, T. (2009). "Risk Assessment for Water Mains using Fuzzy Approach", Building a Sustainable Future. *Proceedings of the 2009 Construction Research Congress*, ASCE, Washington, USA, pp. 1125-1134.

__________. (2010). "Hierarchical Fuzzy Expert System for Risk of Failure of Water Mains". *Journal of Pipeline Systems Engineering and Practice*, 1(1), pp. 53-62.

Infrastructure Report, (2007). "The Water Main Break Clock", Canada Free Press, (Jan. 23, 2008).

Khan. Z., Moselhi. O., and Zayed. T. (2010). "Level of Service Based Methodology for Municipal Infrastructure Management." *International Journal of Human and Social Sciences*, 5(4), pp. 262-271.

Kleiner, Y., Sadiq, R., and Rajani, B. (2004). "Modeling Failure Risk in Buried Pipes using Fuzzy Markov Deterioration Process." *Proc., on Pipeline Engineering and Construction: What's on the Horizon?* ASCE, Reston,VA., pp. 1-12.

Kleiner. Y., Rajani. B., and Sadiq. R., (2006). "Failure Risk Management of Buried Infrastructure using Fuzzy Based Techniques." *Journal of Water Supply: Research and Technology - AQUA*, 55(2), pp. 81-94.

Klir, G.J. and Yuan, B. (2003) *Fuzzy Sets and Logic, Theory and Applications.* Prentice Hall of India Private Limited, New Delhi.

Mandel, J.M. (2001). *Uncertain Rule-based Fuzzy Logic Systems: Introduction and New Directions.* Prentice Hall PTR: New Jersey.

Montgomery, Douglas C. (2000). *Design and Analysis of Experiments* (5th edn.). John Wiley & Sons, Inc.: New York.

Vachhani, Nilesh. (2010). "Determination of Risk for Occurance of Respiratory Diseases among Solid Waste Workers using Fuzzy Rule Based System," M.Tech thesis. Submitted to SVNIT, Surat.

Wang, L.X. (1997). *A Course in Fuzzy Systems and Control.* Prentice-Hall International: New Jersey.

Yan, J. and Vairavamoorthy, K. (2003). Fuzzy Approach for Pipeline Condition Assessment. In *Pipeline Technologies, Security, and Safety*, ASCE, Reston, Va., pp. 467-476.

Zadeh, L.A. (1988). "Fuzzy Logic." *IEEE-CS Computer*, 21(4), pp. 83-93.

Zayed, T. and Fares, H. (2008). "Evaluating Water Main failure Risk Using a Hierarchical Fuzzy Expert System," CSCE 2008 annual conference.

4

Fuzzy Logic Based Operation of Spillway Gates: A Case Study of Ukai Dam, Tapi Basin, India

Utkarsh Nigam[1] and S.M. Yadav

ABSTRACT

Fuzzy logic can be used efficiently to control real-time operation of spillway gates of a reservoir during high inflow or flood. This study presents use of fuzzy logic based operation of spillway gates for high inflow/flood in Ukai reservoir, Tapi basin, India. The real-time data of reservoir has been used to control the operation of spillway gates. In this study high inflow events have been considered and comparison of the observed outflow of dam released with the proposed outflow of fuzzy logic based model has been compared. The proposed method is most systematic as it discharges the water in proportion to the overall severity of the incoming flood hydrograph. The fuzzy control produces smoother, reliable and more absolute outflow hydrographs than those obtained by actual controlled outflow. The results are analysed and justified by comparing the levels of observed and obtained fuzzy elevation which are in good agreement.

Keywords: *Fuzzy logic, gated spillways, real-time operation, reservoir operation.*

INTRODUCTION

The technique of determining the flood hydrograph at a section of a river by utilising the data of flood flow at one or more upstream sections is known as reservoir routing. The range for the water level to be maintained on a particular date has to be prescribed initially. The control system keeps the reservoir water level in prescribed range. This range is called rule level for a reservoir. The nonlinearities occur in reservoir water level flow and these nonlinearities are unexpected. The aim of this fuzzy logic based control system is to adjust the dam elevation as per rule level and to effectively manage the flood during

1. M.Tech, WRE, Department of Civil Engineering, S.V. National Institute of Technology, Surat, Gujarat, India, 395007.

high inflows of short duration by adjusting the openness of spillway gates. The factors that affect reservoir elevation are inflow hydrograph, unexpected and sudden changes in reservoir water level, amount of water discharge per unit of time, maximum possible outflow etc. In this paper method based on fuzzy logic control (FLC) is proposed. Algorithms of fuzzy rules are used to obtain optimized membership function representing fuzzy values.

Operation of a gate closely forms a part to achieve operation of reservoir. Reservoir control were presented initially by Beard (1963) as a deterministic operating procedure for reservoir control. Windsor (1973) tried to form a recursive linear programming procedure for the operation of a flood control system. Can and Houck (1984) gave a goal-programming method for multireservoir system. Ozelkan et al. (1997) applied application of dynamic techniques in reservoir control. Oshimaa and Kosudaa (1998) formed a deterministic chaos method based on the demand prediction in a reservoir control system. Acanal and Haktanir (1999) gave a six-stage operation policy for the routing of flood hydrographs with return periods from 1.01 years up to the Probable Maximum Flood (PMF) for any dam having a gated spillway. Chang and Chang (2001) suggested that optimal hydrological parameters as input and output should be there to build an optimal reservoir operation system. A set of operating rules for 10 stages for controlling the spillway gate opening was made by Haktanir and Kisi (2001). The most common reservoir control strategies are based on maintaining rule level or adhering to water release policies. In this strategy, the amount of water is discharged according to pre-determine rule level. The most significant drawback of this approach is how to determine the constant water amount that must be discharged to maintain a particular level. Another disadvantage of this approach is that it does not consider the rate of change of either the elevation or inflow. Karaboga Dervis et al. (2004) gave a new, reliable and efficient control method based on fuzzy logic which was proposed for the real-time operation of spillway gates of a reservoir during any flood of any magnitude up to the probable maximum flood. To demonstrate the performance of the proposed method the simulation of the control system using different probable overflow hydrographs were carried out by them. Haktanir et al. (2013) proposed a procedure to identify sets of operational rules for gated spillways for optimal flood routing management of artificial reservoirs. They route the flood dividing it into 15 sub-storage, by carrying out a 15 stage routing. The application of fuzzy logic can be traced to the hydrological domain and study by various researchers can found such as by Russel and Campbell (1996), Shrestha et al. (1996), Cheng and Chau (2001), Jolma et al. (2001), Kumar et al. (2001) and Karaboga Dervis (2004). They have applied the fuzzy logic in various trends of reservoir operation. Russel and Campbell (1996) gave fuzzy programming based optimization of reservoir operation and Rule based model by Shrestha et al. (1996). Kumar et al. (2001) presented a method based on fuzzy linear programming for the optimization of the operation policy. An attempt to model the operation of a system of five lakes using a fuzzy logic based approach was made by Jolma et al. (2001).

RESERVOIR MANAGEMENT DURING HIGH INFLOWS

A real-time reservoir operation is a very complex problem and becomes tedious and tough to control for the high inflow for gated spillways. A dam which is built to serve various purposes such as storage, irrigation, power etc. has to be managed effectively by a proper efficient reservoir operation by fulfilling all needs and demands. High inflow or flood usually plays a very important role in the reservoir operation and hence, reservoir control policies have to be adopted to control the outflow to the downstream, to maintain the storage and elevation levels. Satisfying fully and efficiently the demands and needs by mitigating the parameters which affects the purposes of a dam constitute the main objectives of the study. Management of the spillway gates in reservoir operation for gated spillways play a major role during a mild or a severe flood.

Reservoir control and operation has been done by the U.S. Army Corps of Engineers. They utilizes its own method of operating the gated spillways based on "water control diagrams" derived considering long-and short-term hydrologic forecasts (Hydrological 1987). Beard (1963) also presented deterministic rule's approach similar to Corps' method for reservoir operation. Linsley et al. (1992) emphasize that the discharge from a storage reservoir is regulated by gates and valves operated on the basis of the judgement of the project engineer. Sakakima et al. (1992) suggested that for the extremely big flood, a reservoir operator has to control the gates to protect the reservoir and the downstream reference point by relying on his judgement.

FUZZY LOGIC BASED CONTROL SYSTEM

The aim of fuzzy logic based control system is to adjust the dam elevation as per rule level. Various factors that affect reservoir operation are inflow, unexpected and sudden changes in reservoir water level, amount of water discharge per unit of time, maximum possible outflow etc. In this paper gate operation of spillway using fuzzy logic is discussed. Algorithms of fuzzy rules are used to obtain optimized membership function representing fuzzy values. These rules are derived based on the intuition and decision management depending upon the availability of occurrence of particular flow.

The study area is shown in Fig. 1 and Fig. 2 show the schematic diagram for the spillway gate showing parameters in the study. The main variables of a reservoir management system adopted in the present study are the inflow rate [I(t), m^3/s], outflow rate[Q(t), m^3/s], reservoir capacity[S(10^6 m^3)] and minimum reservoir water surface elevation.

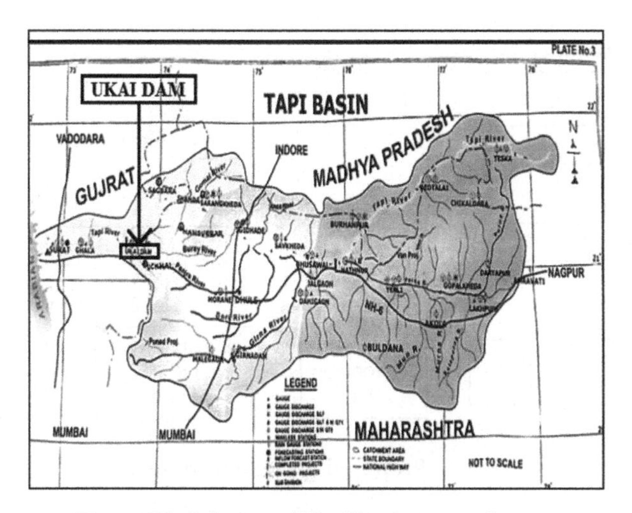

Fig. 1. Ukai dam and Tapi basin as study area

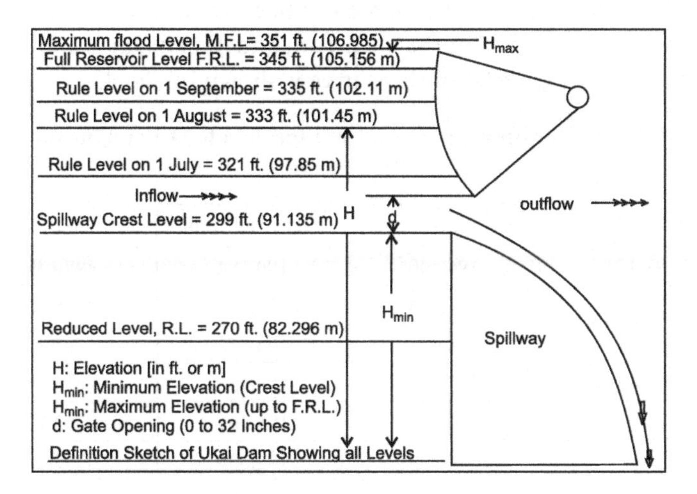

Fig. 2. Schematic diagram of Ukai dam

$H_{\min}$ in metres represent minimum head of water, H represents actual water level in metres and d denotes spillway gate opening in metres. The accumulation of storage in a reservoir depends on the difference between the rates of the inflow and outflow. For the time interval of t, this continuity relationship can be expressed as shown below (Udall 1961):

$$\Delta S(t) = I(t).\Delta t - Q(t).\Delta t$$

Where, $\Delta S(t)$ is storage accumulated or depleted during Δt; and $I(t)/Q(t)$ = average rate of inflow/outflow during Δt.

The fuzzy logic model proposed for the present study is shown in Fig. 3.

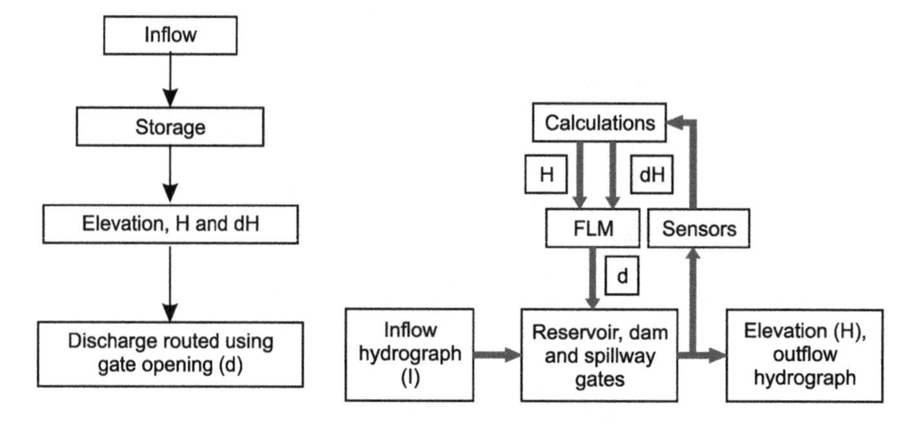

Fig. 3. Proposed fuzzy logic model

Elevation-storage and elevation-discharge curves are developed to find out the outflow hydrograph. The same has been utilized for routing the flow. Elevation and storage data for the Ukai dam and Tapi basin is shown in Table 1. The elevation-storage relationship curve is shown in Fig. 4.

Table 1. Elevation (in m) and storage data for Ukai dam and Tapi basin

Elevation in m	Storage in MCM	Elevation in m	Storage in MCM
91.135	1960.00	100.584	4979.39
91.440	2018.36	102.108	5714.86
92.964	2348.16	103.632	6524.02
94.488	2630.21	105.156	7414.28
96.012	3149.90	105.461	7553.19
97.536	3704.00	106.680	8235.19
99.060	4311.05	106.985	8480.18

To avoid the flooding situations in downstream of a dam, the rule levels are proposed to maintain water levels in the reservoirs. To effectively manage the flood, state authority has revised the rule level during the year 2000 after the flood of 1998. The Ukai reservoir was operated using this rule level till year 2006. In the year 2006, due to flash floods in the catchment area, heavy inflow incurred in the reservoir during short duration (less than 24 hours). According

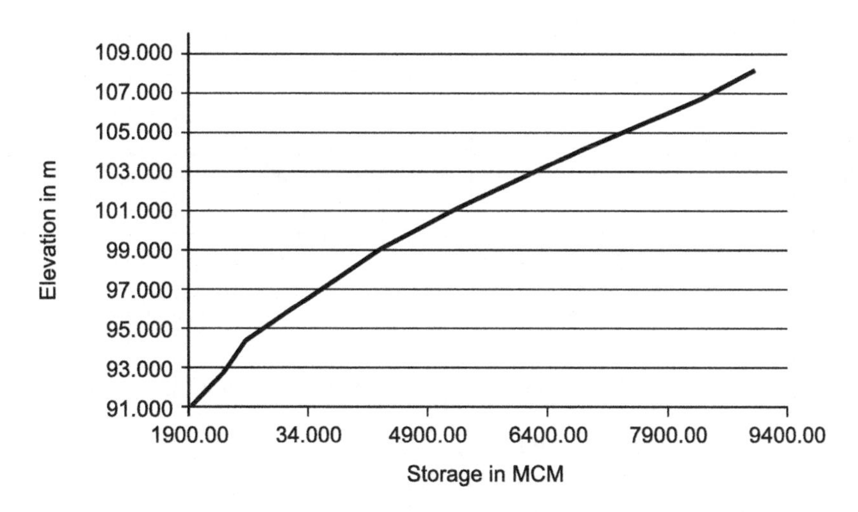

Fig. 4. Elevation (Stage) – Storage curve

to the existing rule level dam was full at that time. The reservoir was unable to absorb heavy inflow of the magnitude around 26350 m^3/s within 24 hours. This forced dam authority to release water from the reservoir. The carrying capacity of the river was approximately 8450 m^3/s and releases of nearly 25788 m^3/s water flooded downstream side. The Surat city which is situated approximately 100 km downstream of the dam and on the confluence of Tapi river and sea remain submerged for three days having depth of water from 0.30 m to 4.5 m. Many people and animal died. The damage to the property and industry was 22,00,000 of rupees. After 2006 flood event rule level of Ukai reservoir was revised. Table 2 represents rule levels after 2000 and 2008. The revised rule level, after 2006 flood event enforced from 2008. Fig. 5 represents the rule level used to maintain reservoir storage for the period 2000–2006 and after 2008.

Table 2. Rule levels for Ukai dam for the year 2000 and 2008

Rule level 2000		Rule level 2008	
Date	R.L. in m	Date	R.L. in m
01-Jul	97.84	01-Jul	97.84
01-Aug	101.50	01-Aug	101.50
01-Sep	104.55	01-Sep	102.11
16-Sep	104.85	16-Sep	103.63
01-Oct	105.16	01-Oct	105.16

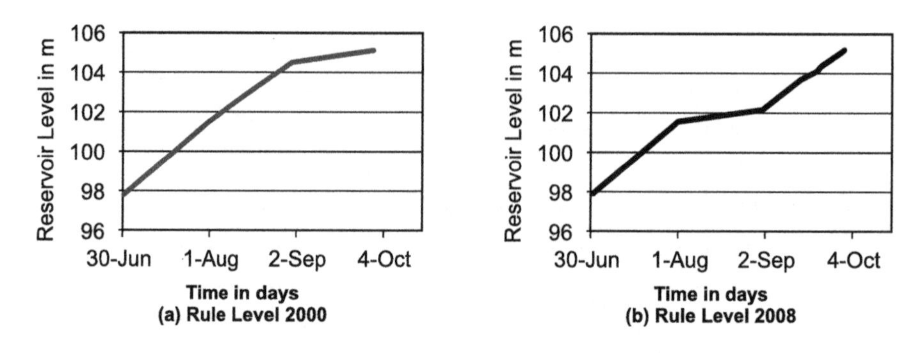

Fig. 5. Rule levels for period 2000-2008 and after 2008 for Ukai reservoir

These rule levels are used to develop the fuzzy logic based controlled system. As per the rule level to be maintained at the beginning of the month is taken as an initial limit for the development of model. For fuzzy logic based model the input and required output parameters are discussed as under.

Input and Output Variables

The input variables for the fuzzy model are lake level (*H*) and rate of change in Elevation (*dH*). The output of the model is gate opening (*d*). The outflow rate of the reservoir is controlled by the gate opening tuned by the Fuzzy Logic model. For the *H*, *dH* and *d* variables, the normalization interval are shown below in Table 3.

Table 3. Normalization interval of the parameters used

Parameters	Range	
	Maximum	Minimum
Elevation (*H*)	Maximum Flood Level, M.F.L. (106.98m)	Rule Level
Rate of change in elevation (*dH*)	+1	–1
Gate opening (*d*)	0.8128 m (81.28 cm)	0

The ranges of inputs are based on the analysis of previous 15 years past data of elevation, inflow, change in inflow and releases. The gate opening ranges i.e. output are made by users intuition by analyzing data of gate opened after flood event of 2006 to 2013.

Crest level of reservoir is: 91.135 m, Full Reservoir Level (F.R.L.) is 105.148 m and Maximum Flood Level (M.F.L.) is 106.99 m.

For single purpose reservoir only one model is required but here the reservoir is multipurpose and a pre-fixed elevation known as "Rule Level" is required to be maintained. The highest inflow in the reservoir is likely to occur during July to September. For monsoon months operating strategies for reservoir operation are proposed through fuzzy logic technique.

The inflow, releases and storage data were collected for 15 years period and gate opening datas were collected for 7 years (2006 to 2012). After analyzing the data the range of elevation/lake level for monsoon months are tabulated in Table 4.

Table 4. Range of elevation/lake level

Month	Elevation/lake level (in m)	
	Maximum	Minimum
July	106.985	97.84
August	106.985	100.889
September	106.985	102.108

Membership Functions

The membership functions used for the fuzzy values of fuzzy variables are selected based on human/expert experience. The fuzzy values are represented by triangular/trapezoidal membership functions for the present study. In the present study five membership functions for inputs Elevation (*H*) and change in Elevation (*dH*) and output Gate opening (*d*) are used.

For "*H*" five membership functions used are: very low, low, medium, high and very high. For "*dH*" five membership functions used are: negative big, negative small, zero, positive small and positive big. Lastly for "*d*" (gate opening) five membership functions used are: very low, low, medium, high and very high. The ranges of membership functions are shown in Table 5. The structure of rules are then formed to derive the relationship between input and output then finally processed to get output. Mamdani's maximum-minimum method is used in present study for fuzzy inference and standard centre of area (centroid) method is used for the defuzzification of output. Figure 6 shows the membership functions used in fuzzy logic based model.

Table 5. Parameters used in fuzzy logic based model (Input, output, membership function details)

S. No.	Parameters	
1.	Input parameter	*H* (Elevation), *dH* (change in elevation)
2.	Output parameters	*d* (gate opening)
3.	Membership function for each parameter *H* *dH* *d*	5 5 5
4.	Total number of rules derived for programme, in fuzzy logic tool	25
5.	Membership function used	Triangular / Trapezoidal membership function
6.	Inference mechanism	Mamdami type inference mechanism
7.	Spillway crest	91.135 m (299 ft.)
8.	Highest flood level	106.990 m (351 ft.)
9.	Full reservoir level	105.156 m (345 ft.)

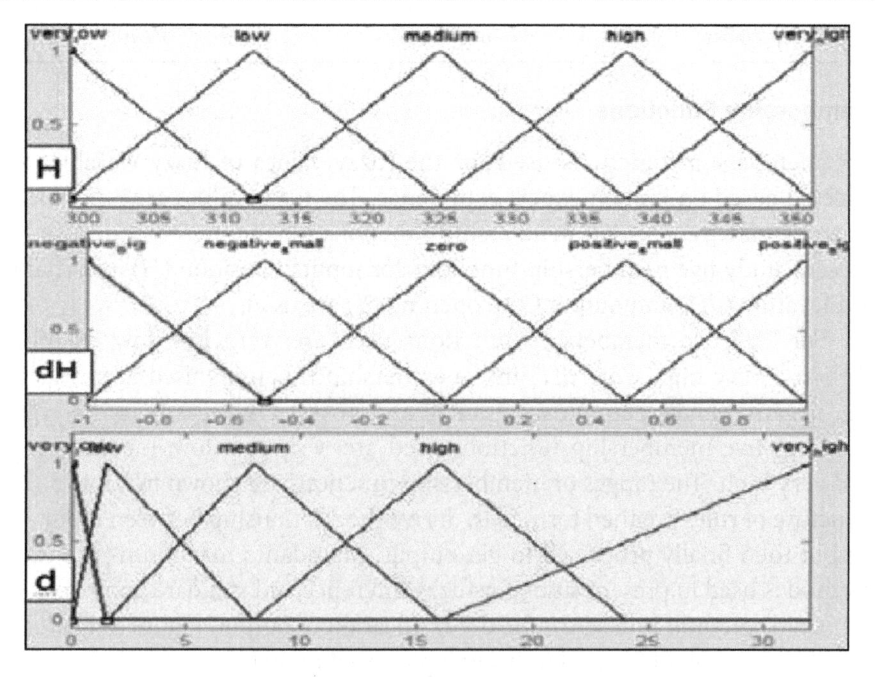

Fig. 6. Membership functions used in fuzzy based model

Structure of Rules

The rules of the fuzzy logic based model are derived from the analysis of past data of reservoir operation. The rule base of the fuzzy logic based model is shown in Table 7.

Table 7. Relation developed for fuzzy Logic programme between membership function

→dH H↓	Negative less	Negative big	Zero	Positive less	Positive big
Very less	Very less	Very less	Very less	Very less	Very less
Less	Less	Less	Less	Less	Less
Medium	Medium	Medium	Medium	Medium	Medium
Large	Large	Large	Large	Large	Large
Very large	Very large	Very large	Very large	Very large	Very large

Some examples of fuzzy rules are:

- If dam Elevation is less and rate of change of Elevation is less positive, then the spillway gate opening is very less.
- If dam Elevation is at medium and rate of change of Elevation is zero, then the spillway gate opening is very less.
- If dam Elevation is large and rate of change of Elevation is less positive, then the spillway gate opening is at medium.
- If dam Elevation is very large and rate of change of Elevation is less negative, then the spillway gate opening is less.

In this study, the real data of flood hydrograph of Ukai dam are used and simulations are carried out for five actual inflow hydrographs as shown in Fig. 7.

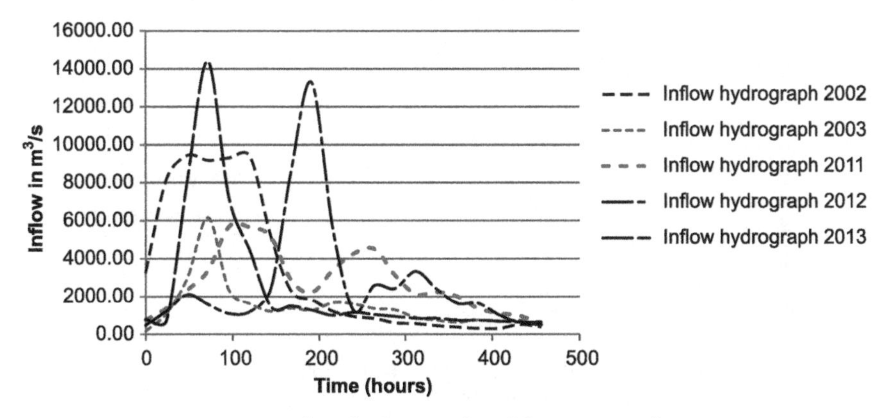

Fig. 7. Inflow hydrographs of five years inflow

The initial value of the reservoir elevation is assumed as 91.135 m which is given as top of spillway. The full reservoir level is 105.156 m. For Ukai dam the relation between dam capacity and lake level is given above in Table 1.

The simulation results produced using fuzzy logic based model for five inflows are given in Figs. 8 to 12.

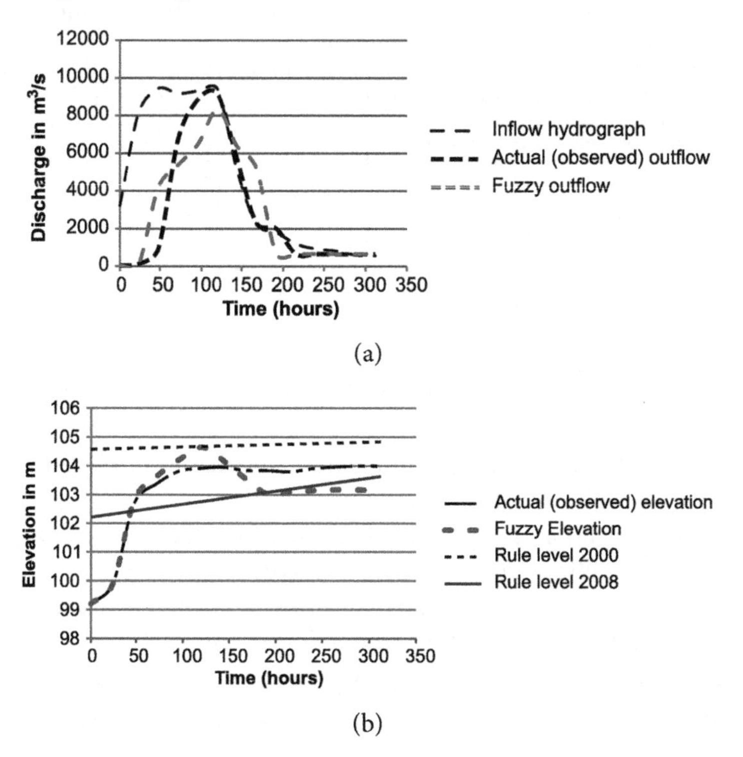

Fig. 8. Performance characteristics of fuzzy logic based model for year 2002 ($I_{max.}$ = 9391.37 m^3/s)

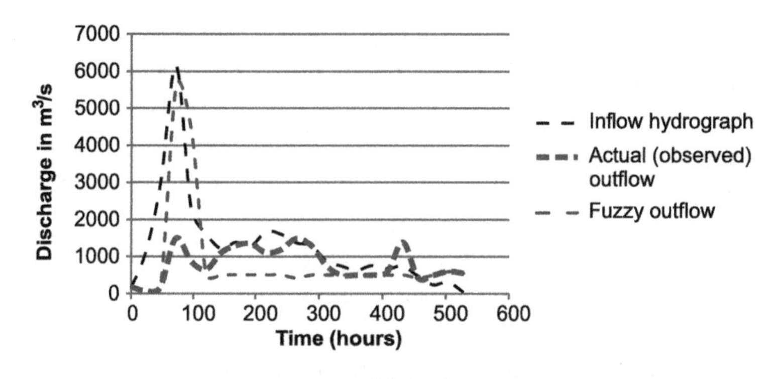

(a)

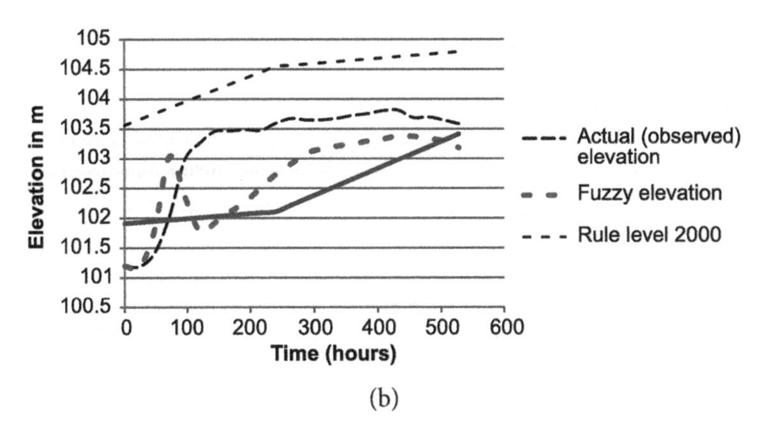

(b)

Fig. 9. Performance characteristics of fuzzy logic based model for year 2003 ($I_{max.}$ = 6152.25 m^3/s)

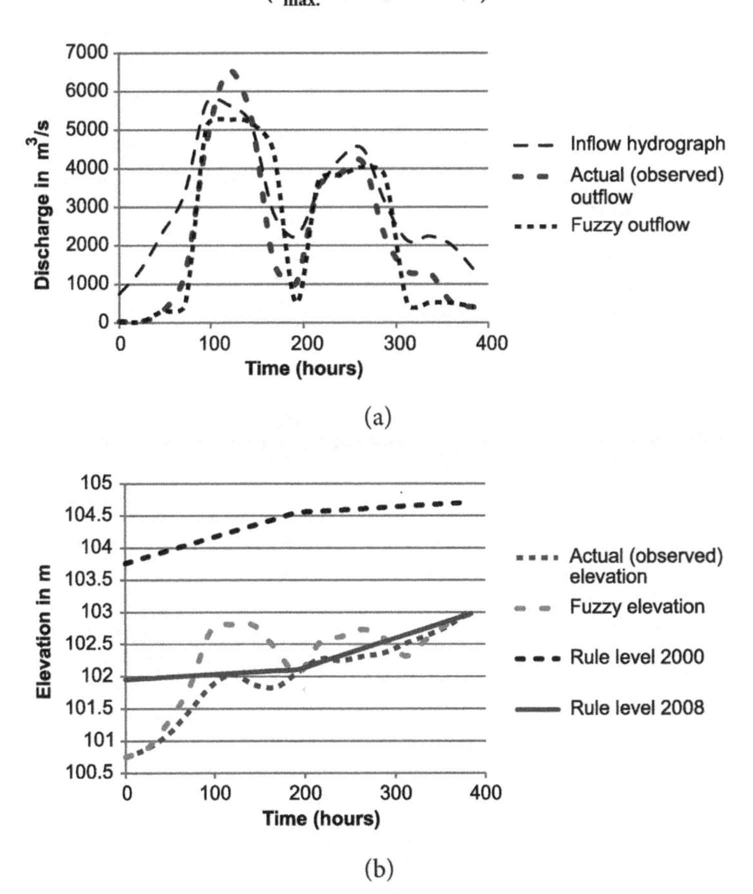

(b)

Fig. 10. Performance characteristics of fuzzy logic based model for year 2011 ($I_{max.}$ = 5691.30 m^3/s)

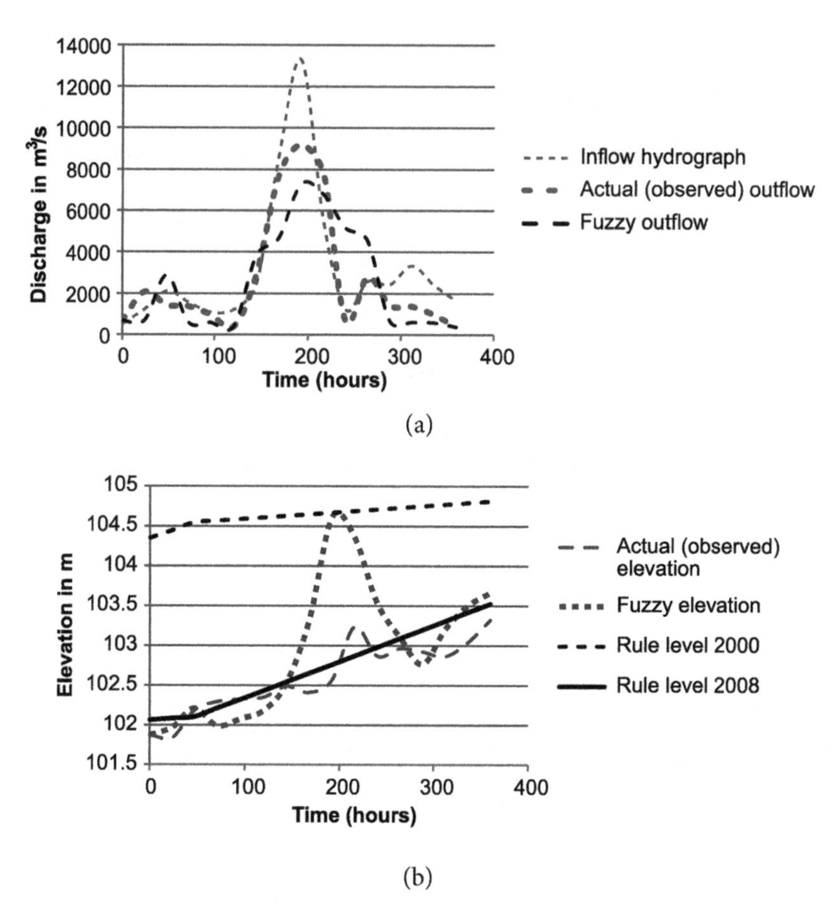

(a)

(b)

Fig. 11. Performance characteristics of fuzzy logic based model for year 2012 ($I_{max.}$= 13276.08 m^3/s)

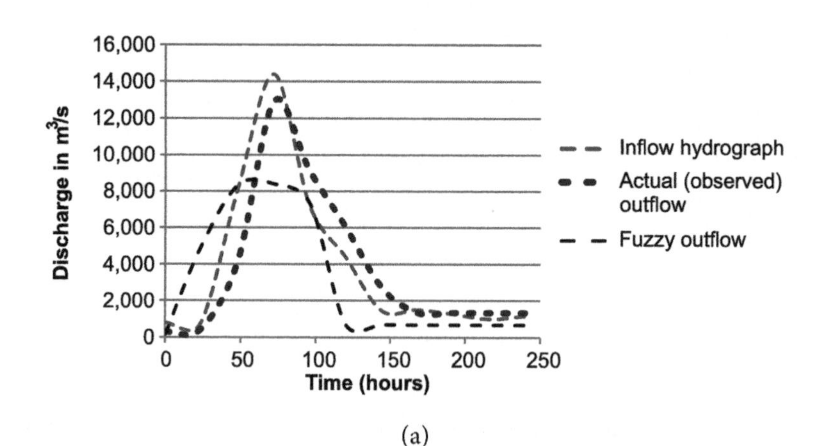

(a)

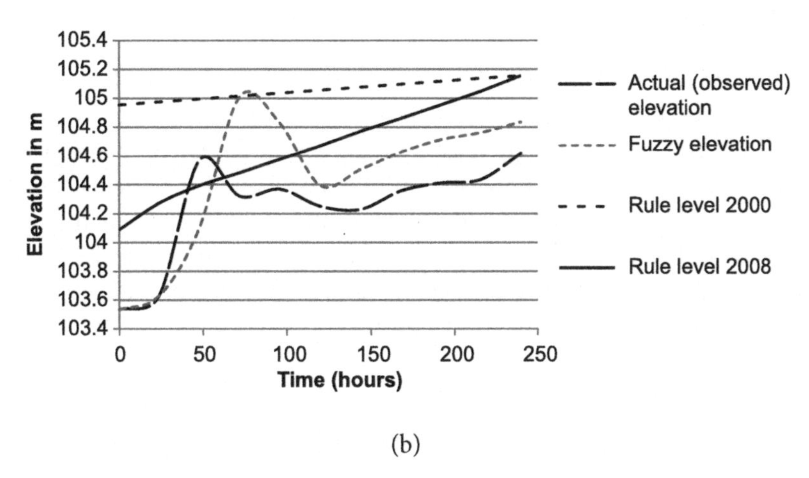

(b)

Fig. 12. Performance characteristics of fuzzy logic based model for year 2013 ($I_{max.}$= 14391 m³/s)

Flood or high inflow can be mitigated by using fuzzy logic based model without manual interference and hence protects the flood at the downstream side of dam. Surat city (India) is situated 100 kms downstream to Ukai dam, at the tail end of river Tapi is subjected to moderate to heavy floods frequently due to heavy rainfall in the catchment area. To minimize frequent flooding of city Surat, certain preventive measures in terms of controlled release from the dam can be imposed. The carrying capacity of Tapi river at Surat is about 8490 m^3/s. In the present study outflow from Ukai dam is restricted to release flow up to 8490 m^3/s. Fuzzy logic based model has been developed to mitigate the flood, to minimize or reduce the expected damages and casualties on the downstream side of Ukai dam.

From the above simulation results obtained by using real data, the following conclusions can be drawn:

1. The fuzzy logic based operation can be carried out automatically without requiring any human operator interference.
2. The proposed control method is the most systematic approach, since it discharges the water in proportion to the overall severity of the incoming flood hydrograph.
3. The fuzzy control produces smoother, reliable and more absolute outflow hydrographs than those obtained by actual (observed) controlled outflow.
4. The fuzzy control successfully decreases the elevation/lake level of the reservoir to the desired level for a high inflow hydrograph.

CONCLUSION

In this paper a fuzzy logic based new, advanced, effective and efficient control method is presented for the real-time flood operation of spillway gates of Ukai dam. The results of fuzzy logic based proposed model is compared with the actual (observed) values of downstream releases. Simulation results show that the fuzzy logic based model is more flexible and desirable results than actual (observed) outflow. The characteristics of the fuzzy logic control system are highly desirable in reservoir management, and are important indications of the power and effectiveness of the fuzzy control approach.

REFERENCES

Acanal, Nese and Haktanir, Tefaruk. (1999) "Six-Stage Flood Routing for Dams Having Gated Spillways". *Turkish Journal of Engineering and Environmental Science,* 23, pp. 411- 422.

Acanal Nese, Yurtal, Recep and Haktanir, Tefaruk. (2000). "Multi-Stage Flood Routing for Gated Reservoirs and Conjunctive Optimization of Hydroelectricity Income with Flood Losses". *Hydrological Sciences-Journal-des Sciences Hydrologiques,* 45(5), October, pp. 675-688.

Beard, L.R. (1963). "Flood Control Operation of Reservoirs." *Journal of the Hydraulics Division, American Society of Civil Engineers,* 89(1), pp. 1-23.

Can, E.K. and Houck, M.H. (1984). "Real-Time Reservoir Operations by Goal Programming." *Journal of Water Resources Planning and Management,* 110(3), pp. 297-309.

Chang, L. and Chang, F. (2001). "Intelligent Control for Modeling of Real-Time Reservoir Operation." *Hydrological Processes,* 15, pp. 1621-1634.

Davis, Calif. (1987). "Management of Water Control System, Engineering and Design." Hydrological Engineering Center, Rep. EM 1110-2-3600, U.S. Army Corps of Engineers.

Haktanir, T. and Kisi, Ö. (2001). "Ten-Stage Discrete Flood Routing for Dams Having Gated Spillways." *Journal of Hydrologic Engineering,* 6(1), pp. 86-90.

Haktanir, Tefaruk, Citakoglu, Hatice and Acanal, Nese. (2013). Fifteen-Stage Operation of Gated Spillways for Flood Routing Management through Artificial Reservoirs. *Hydrological Sciences Journal - Journal des Sciences Hydrologiques,* 58(5), pp. 1013-1031.

Jolma, A., Trunen, E., Kummu, M. and Dubrovin, T. (2001). "Reservoir Operation by Fuzzy Reasoning." *Proceeding International Congress on Modeling and Simulation,* MODSIM, pp. 10-13.

Karaboga Dervis, Bagis Aytekin and Haktanir Tefaruk (2004). "Fuzzy Logic Based Operation of Spillway Gates of Reservoirs during Floods". *Journal of Hydrologic Engineering,* ASCE, 9(6), pp. 544-549.

Klir, George J. and Yuan, Bo (1995). *Fuzzy Sets and Fuzzy Logic: Theory and Applications.* Prentice Hall, PTR: Upper Saddle River New Jersey.

Kisi, Ö. (1999). "Optimum Ten Stage Overflow Operating Model for Dams Having Gated Spillway." M.Sc. thesis, Erciyes University, Turkey.

Kumar, D.N., Prasad, D.S.V. and Raju, K.S. (2001). "Optimal Reservoir Operation using Fuzzy Approach." *Proceedings of International Conference on Civil Engineering (ICCE)*, 2, pp. 377-384.

Linsley, R.K., Franzini, J.B., Freyberg, D.L. and Tchobanoglous, G. (1992). *Water Resources Engineering*, 4th edn., McGraw-Hill: New York.

Oshimaa, N., and Kosudaa, T. (1998). "Distribution Reservoir Control with Demand Prediction using Deterministic-Chaos Method." *Water Science and Technology,* 37(12), pp. 389-395.

Ozelkan E.C., Galambosia, Á., Gaucheranda, E.F. and Duckstein, L. (1997). "Linear Quadratic Dynamic Programming for Water Reservoir Management." *Applied Mathematical Modelling,* 21(9), pp. 591-598.

Russell, Samuel O. and Campbell, Paul F. (1996). "Reservoir Operating Rules with Fuzzy Programming". *Journal of Water Resources Planning and Management,* May/June.

Sakakima, S., Kojiri, T., and Itoh, K. (1992). "Real-Time Reservoir Operation with Neural Nets Concept." *Proc., 17th Int. Conf. on Applications of Artificial Intelligence in Engineering—AIENG/92,* Computational Mechanics Publications, Southampton, U.K., pp. 501-514.

Shrestha, B.P., Duckstein, L. and Stakhiv, E.Z. (1996). "Fuzzy Rule Based Modeling of Reservoir Operation." *Journal of Water Resources Planning And Management,* 122(4), pp. 262-269.

Udall, S.L. (1961). "Design of Small Dams." *Rep.,* U.S. Dept. of the Interior, Bureau of Reclamation, Washington, D.C.

Windsor, J.S. (1973). "Optimization Model for the Operation of Flood Control Systems." *Water Resources Research,* 9(5), pp. 1219-1226.

Wurbs, R.A. (1993). "Reservoir System Simulation and Optimization Models." *Journal of Water Resources Planning and Management,* 119(4), pp. 455-472.

Yeh, W. (1985). "Reservoir Management and Operations Models: A State of the Art Review." *Water Resources Res*earch, 21(12), pp. 1797-1818.

5

Groundwater Quality Mapping and Predictions Using Geospatial and Soft Computing Techniques

Raj Mohan Singh[1] and Prachi Singh[2]

ABSTRACT

Groundwater is useful for irrigation and drinking water in both urban as well as in rural area. However, groundwater is vulnerable to pollution and depletion, and needs to be managed carefully to avoid serious economic and environmental costs. Contamination of groundwater poses serious threat to the environment. The contamination of aquifer not only threatens public health and the environment, it also involves large amounts of money in fines, lawsuits and cleanup costs. Once groundwater is contaminated, it may be difficult and expensive to clean up. Sometimes it is almost impossible to clean it up to drinking water standards. Therefore, it is imperative to know the spatial and temporal extent of the pollution so that necessary remedial steps may be taken. Continuous monitoring of water quality is costly affair. Also, water managers need future predictions for water quality parameters. Prediction model based on soft computing techniques (fuzzy system and ANN) incorporate system uncertainty in predictions.

In this work the groundwater quality distribution mapping is performed using geospatial techniques. Both variations (spatially and temporally) were mapped using arcGIS. The observation data for 6 water quality parameters i.e total dissolved solids (TDS), electrical conductivity (EC), calcium ion (Ca^{2+}), sodium ion (Na^{+}), magnesium ion (Mg^{+2}), nitrate ion (NO_3^-) for 22 locations in four different seasons are used for spatiotemporal representation using different kriging techniques in the study area. Correlation analysis was performed to assess the salinity represented by electrical conductivity. Correlation parameters are then further used for prediction of salinity. Regression, ANN and fuzzy techniques are used for water quality prediction. Performance of the models were evaluated using some performance indices like average percent error (APE), percent bias (PBIAS), Nash-Sutcliffe efficiency (NSE), root mean square error (RMSE).

Keywords: *Groundwater, spatiotemporal, regression, ANN, fuzzy*

1. Associate Professor, Department of Civil Engineering, Motilal Nehru National Indian Institute of Technology, Allahabad, 211004, India, Email: rajm.mnnit@gmail.com, rajm@mnnit.ac.in
2. Research Scholar, Department of Civil Engineering, Motilal Nehru National Indian Institute of Technology, Allahabad, 211004, India, Email: prachichandel272@gmail.com

INTRODUCTION

Groundwater is the major source of irrigation and drinking water in both urban and rural India. Even in those areas where surface water is abundant, the groundwater is used as practical source of water for public supply, agriculture and industry due to its quality, easy accessibility, reliability and relatively low cost associated with its use. However, groundwater is vulnerable to pollution and depletion, and needs to be managed carefully to avoid serious economic and environmental costs.

The assessment of spatial correlation in hydrochemical variables is an important tool in the analysis of groundwater chemistry especially in arid and semiarid zones. In geostatistical methods, kriging technique is quite popular. The technique is applicable to cases such as determining groundwater level, estimation of hydrochemical distribution of soil properties, and other estimations. Kumar and Remadevi (2006) have compared various variogram models, i.e., spherical, exponential, and Gaussian semivariogram models to fit the experimental semivariograms in identifying the spatial analysis of groundwater levels in small part of Rajasthan area, India. The results indicate that the kriged groundwater levels satisfactorily matched the observed groundwater levels. Ahmadi and Sedghamiz (2007) used ordinary and universal kriging methods in identifying the spatial and temporal analysis of monthly groundwater level fluctuations in the arid region, the Darab Plain, in the south of Iran. Results revealed that the spatial and temporal variations of groundwater level and groundwater level fluctuations were underestimated by 3 and 6 %, respectively. Very low variation and acceptable errors support the unbiased hypothesis of kriging. Shamsudduha (2007) used different interpolation procedures in evaluating the most appropriate prediction method for the estimation of arsenic concentrations in the shallow aquifer in Bangladesh. In this case, ordinary kriging on original arsenic concentrations and their residual values produced better prediction models regarding the cross-validation, mean prediction error and biasedness analyses.

The role of geographic information system (GIS) software in analyzing spatial distribution of groundwater has been investigated by many authors such as Mehrjardi et al. (2008). Pradhan (2009) studied groundwater potential zonation for basaltic watersheds. Ayazi et al. (2010) investigated disasters and risk reduction in groundwater. Manap et al. (2012) applied the probabilistic based frequency ratio in groundwater potential. El Afly (2012) and Machida et al. (2012) applied an integrated Geostatistics and GIS technique in groundwater. Neshat et al. (2013) estimated groundwater vulnerability to pollution, and Manap et al. (2013) studied prediction of groundwater potential zones.

The objectives of the present research are to investigate the application of various spatial models to interpret the spatial distribution of the groundwater quality and to predict the general trend of the spatial distribution of groundwater in the recharge area of Muzaffarnagar district, Uttar Pradesh in Yamuna-Krishni sub-basin. In this study, kriging techniques in the framework of GIS software (ArcGIS Geostatistical Analyst) are used. Water quality parameters further used for water quality prediction using regression, ANN and fuzzy methods. Models were evaluated on the basis of performance indices.

DESCRIPTION OF THE STUDY AREA AND DATA COLLECTION

The study area falls in the district of Muzaffarnagar between longitudes 77°05′ and 77°27′ E and latitude 29°15′ and 29°41′ N and covers an area about 1100 km^2. It lies in the interfluves of Ganga and Yamuna rivers in the western most part of the Uttar Pradesh. The Rivers Yamuna and Krishni forms the western and eastern boundaries, respectively. The study area is being famous as an intensive agriculture tract of western Uttar Pradesh. Heavy withdrawal of groundwater has set a declining trend of water table over the decade. Few blocks of the basin is reported to be over exploited and some are in semi-critical to critical position. With rise in population and agricultural development, the withdrawal will go at higher scale, which needs special study in order to ascertain the future behaviour of water table in time and space. The Eastern Yamuna Canal and its distributaries are main source of surface water irrigation (CGWB, 2008).

Water quality data is available for 22 wells collected in the four different seasons. The observation data for 6 are used for spatiotemporal representation using different kriging techniques. Correlation analysis was performed to assess the salinity represented by electrical conductivity. Correlation parameters are then further used for prediction of salinity represented by electrical conductivity. Both regression and fuzzy techniques are used for prediction. The performance evaluations of the prediction methodology in terms of specified performance criteria are encouraged.

METHOD OF SPATIOTEMPORAL ANALYSIS

Spatial distribution of groundwater chemistry parameters is analysed using GIS software. In this research, classification of concentration values to identify water quality for drinking water purposes follows the approach by McNeely et al. (1979). Flow chart showing methodologies adopted are:

1. Base map of study area
2. Georefrencing or geometric correction in the base map
3. Digitization of map
4. Assigning attributes for the study area
5. Interpolation using different ordinary kriging methods
6. Representation of contour map

Geostatistical analysis is used to model the spatial distribution of groundwater chemistry. Geostatistics can be regarded as a collection of numerical techniques that deal with the characterization of spatial attributes employing primarily random models in a manner similar to the way in which time series analysis characterizes temporal data (Olea 1999). It deals with spatially autocorrelated data that have a basic structure or spatial patterns which can be manifested in (semi)variogram analysis. (Semi)variogram is a characterization of the spatial correlation of the variables under study. The semivariogram shows the relationship between the lag distance on the horizontal axis and the semivariogram value on the vertical axis. Lag distance is the distance between the measurements of a particular property. From the semivariogram, the spatial correlation of a spatially varying property can be described. The semivariogram value increases from low to high values indicating higher spatial autocorrelation at the small lag distance (Nayanaka et al. 2010).

Kriging is a technique of making optimal, unbiased estimates of regionalized variables at unsampled locations using the structural properties of the semivariogram and the initial set of data values (David 1977). The general equation of the kriging method has been discussed in details in the Marko et al. 2014.

FUZZY RULE BASE SYSTEM

Fuzzy inference system has been used in this work for flow prediction. FIS is a soft computing technique primarily used for analyzing complex systems, especially when the data structure is characterized by several linguistic parameters. These linguistic parameters have relationship with fuzzy concepts like fuzzy sets, rule base, linguistic (approximate) variables etc., which account for the effect upon the output of a system due to various input values, without the need for a definite threshold value (Mitra et al., 2008). Different models in this work are developed on MATLAB FIS editor. The entire FIS framework is based on the concept of fuzzy set theory, fuzzy IF–THEN rules and fuzzy reasoning (defuzzification). FIS can utilize human expertise by storing its essential components in rule base and perform fuzzy reasoning to infer the overall output value (Nash and Sutcliffe, 1970). The theory of fuzzy logic or soft computing was introduced by Lotfi A. Zadeh (1965) to model the uncertainty of the natural language. FIS acquires knowledge from domain experts and that is encoded within the algorithm in terms of the set of IF–THEN rules and employs this rule-based approach and interpolative reasoning to respond to new inputs (Mitra et al., 2008). Fuzzy inference system has four components knowledge base, fuzzification interface, decision making unit and the defuzzification interface.

In the fuzzification interface the input parameters are fuzzified by assigning suitable membership functions to each input parameter. The shapes of the input parameter membership function used in this work are triangular and trapezium represented as trimf and trapmf in the MATLAB FIS editor. Further these data is used in the processing and finally the output is defuzzified to give the output.

ARTIFICIAL NEURAL NETWORK (ANN)

ANN is another kind of soft computing technique based on analogy with brains of living systems, and work as parallel distributed information processing systems. A neural network consists of neurons, called nodes, connected by weighted edges (which represent synapses). Typically, a neural network will have an input and an output layer, and one or more *hidden layers* in between. The neurons in the same layer are not connected to one another, but only to the preceding and following layers. More complicated systems, permitting feedback, are also possible. The weights of a network's nodes and synapses are set during a *training phase.* During training, inputs are presented to the network, and feedback from its outputs used to correct its internal weights. Over time, these weights adjust so that the correct output is produced for all, or most, inputs. A trained network is then presented with an input it has never seen; these weights are expected to produce the right output.

The ANN design called multilayer perceptron (MLP) is especially suitable for classification and is widely used in practice. The network consists of one input layer, one or more hidden layers and one output layer, each consisting of several neurons. Each neuron processes its inputs and generates one output value that is transmitted to the neurons in the subsequent layer. Each neuron in the input layer (indexed $i = 1, \ldots, n$) delivers the value of one predictor (or the characteristics) from vector x. When considering default/non-default discrimination, one output neuron is satisfactory. In each layer, the signal propagation is accomplished as follows: a weighted sum of inputs is calculated at each neuron: the output value of each neuron in the preceding network layer times the respective weight of the connection with that neuron. There are two stages of optimization. First, weights have to be initialized, and second, a nonlinear optimization scheme is implemented. In the first stage, the weights are usually initialized with some small random number. The second stage is called learning or training of ANN.

RESULT AND DISCUSSION

Results for Spatiotemporal Analysis

The spatial distribution of TDS, EC, Ca, Na, NO_3 and pH for four different seasons and five different kriging (spherical, circular, exponential, Gaussian and linear) are performed to generate different interpolation maps. From the various kriging techniques used in the spatial distribution, it is clear that spherical kriging gives much accurate and clear distribution pattern as compared to other kriging method. Here spatial distribution of TDS and EC for June 2007 using spherical kriging techniques are shown in Figs. 1 and 2.

Water Quality Prediction

Correlation Matrix

The correlation matrix for pre-monsoon and post-monsoon are shown in Tables 1 and 2 respectively.

Table1. Correlation matrix of groundwater quality parameters of pre-monsoon

	EC	TDS	AMSL	Mg	Ca	Na	K	NO_3	Hardness	HCO_3	Cl	SO_4
EC	1											
TDS	.717	1										
AMSL	-.287	-.275	1									
Mg	.422	.437	-.060	1								
Ca	.099	.096	.123	.097	1							
Na	.604	.893	-.351	.149	-.038	1						
K	.258	.190	.063	.321	.023	-.083	1					
NO_3	.393	.434	-.104	.372	.143	.326	.160	1				
Hardness	.344	.387	-.028	.913	.407	.079	.290	.342*	1			
HCO_3	.252	.581	-.253	.015	-.278	.571	.029	-.216	-.054	1		
Cl	.606	.578	-.135	.532	.107	.452	.370	.272*	.493	.182	1	
SO_4	.516	.590	-.012	.390	.415	.451	.092	.498	.425	-.205	.229	1

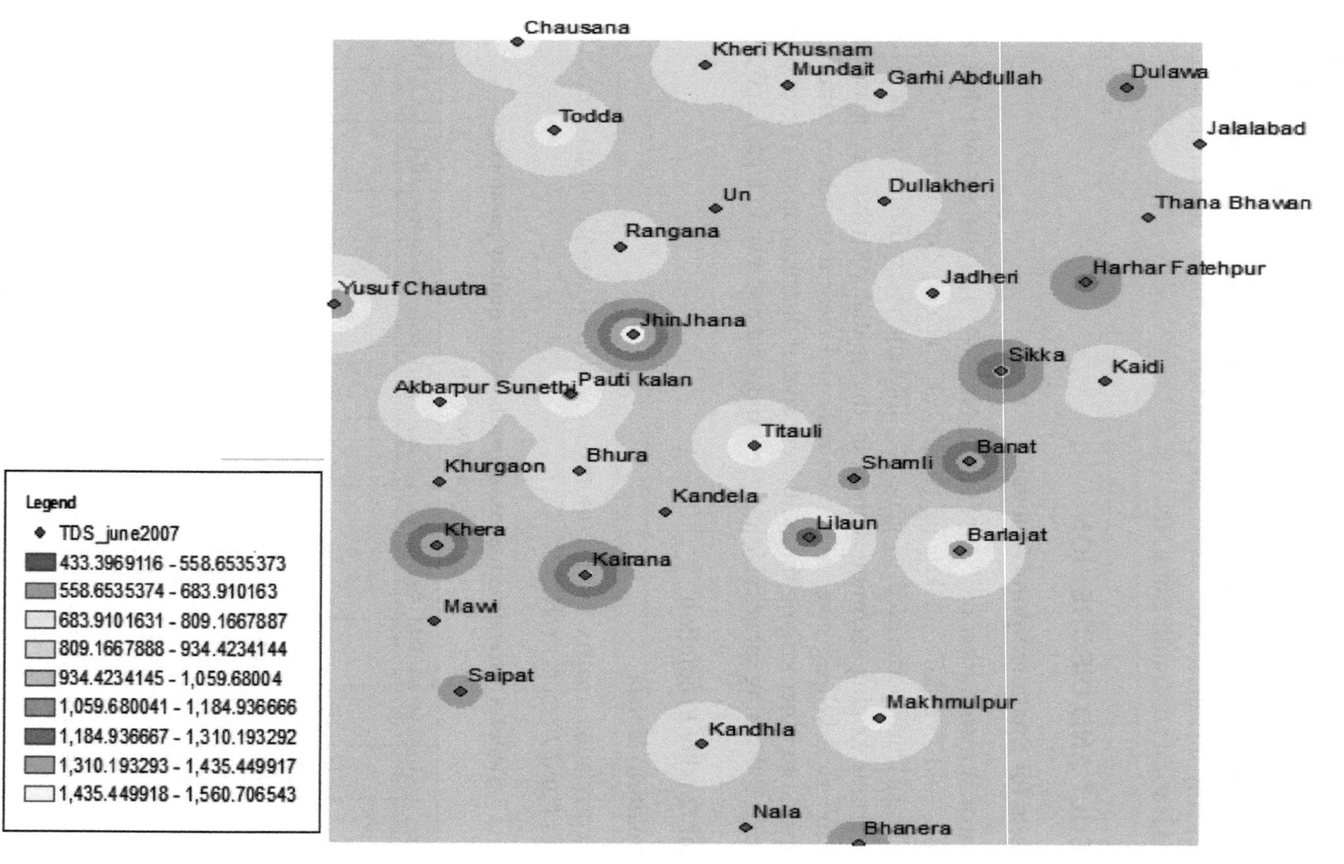

Fig. 1. Interpolation of TDS using spherical kriging in June 2007

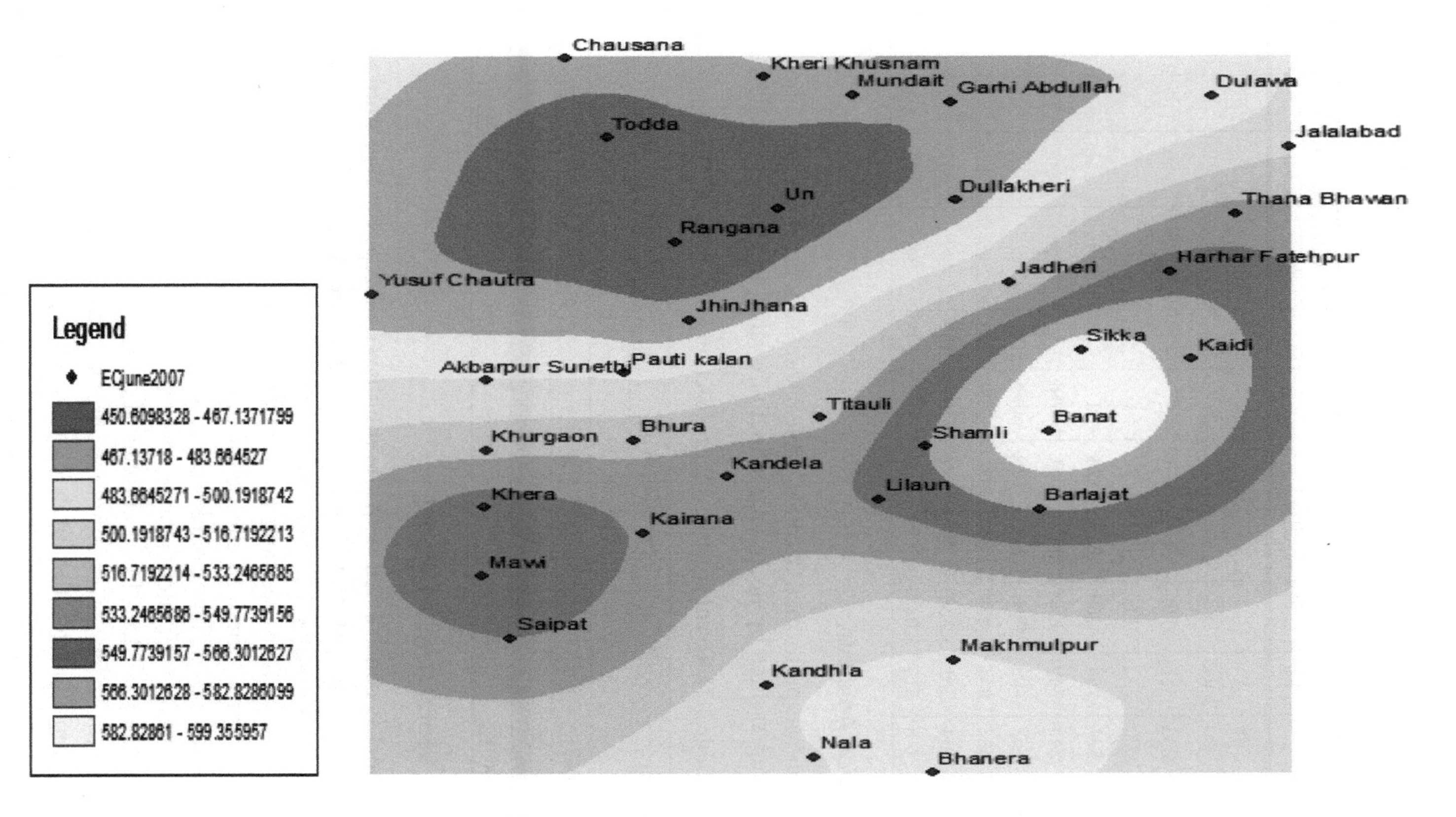

Fig. 2. Interpolation of EC using spherical kriging in June 2007

Table 2. Correlation matrix of groundwater quality parameters of post-monsoon

	EC	TDS	AMSL	Mg	Ca	Na	K	Hardness	HCO_3	Cl	SO_4
EC	1										
TDS	.708	1									
AMSL	-.191	-.399	1								
Mg	.232	.316	.020	1							
Ca	.070	.272	-.117	-.179	1						
Na	.531	.764	-.460	-.077	.126	1					
K	.245	.249	.172	.075	.061	-.018	1				
Hardness	.262	.408	.004	.940	.129	-.053	.196	1			
HCO_3	.454	.834	-.418	.151	.320	.597	.022	.246	1		
Cl	.534	.361	-.150	.538	-.146	.198	.105	.469	.041	1	
SO_4	.466	.514	-.042	.253	.080	.328	.242	.288	.088	.215	1

Regression Analysis

Regression was performed in SPSS. Table 3 shows the regression equations for salinity (represented by electrical conductivity value, EC) using only TDS.

Table 3. Different equations for prediction of EC using TDS only

Type of regression	Pre-monsoon	Post-monsoon
Linear	EC = 0.856TDS-285.13	EC = 0.54TDS-64.86
Exponential	$EC = 145.1exp^{0.001TDS}$	$EC = 181.1exp^{0.001TDS}$
Power	$EC = 0.017TDS^{1.5}$	$EC = 0.234TDS^{1.3}$
Polynomial	EC = 0.317TDS-15.26	EC = 1.01TDS-304.84

Artificial Neural Network Model

ANN model is constructed with the multilayer perceptron algorithm. Both pre-monsoon and post-monsoon season model have an architecture of 1-2-1 neural network, means that there are total 1 independent variables (input), 2 neurons in the hidden layer and 1 dependent (output) variable. SPSS software is used. SPSS procedure can choose the best architecture automatically and it builds the network with one hidden layer. It is also possible to specify the range of units allowed in the hidden layer, and the automatic architecture selection procedure finds out the "best" number of units (2 units are selected for this analysis) in the hidden layer. Automatic architecture selection uses the default activation functions for the hidden layer (Hyperbolic Tangent) and output layers (softmax). Further, 70% of the data is allocated for the training (training sample) of the network and to obtain a model; and 30% is assigned as testing sample to keep tracks of the errors and to protect from the overtraining. Different types of training methods are available like batch, online and minibatch. Here, batch training is chosen because it directly minimizes the total error and it is most useful for "smaller" datasets. Moreover, optimization algorithm is used to estimate the synaptic weights and "Scaled Conjugate Gradient" optimization algorithm is assigned because of the selection of the batch training method. Batch training method supports only this algorithm. Additionally, stopping rules are used to determine the stopping criteria for the network training. According to the rule definitions, a step corresponds to iteration for the batch training method. Here, one maximum step is allowed if the error is not decreased further. Here, it is important to note that, to replicate the neural network results exactly, data analyser needs to use the same initialization value for the random number generator, the same data order and the same variable order, in addition to using the same procedure settings.

The ANN structure for pre-monsoon and post-monsoon season has been shown in the Figs. 3 and 4. The darker lines indicate synaptic weight less than 0 whereas lighter lines indicate synaptic weight greater than 0.

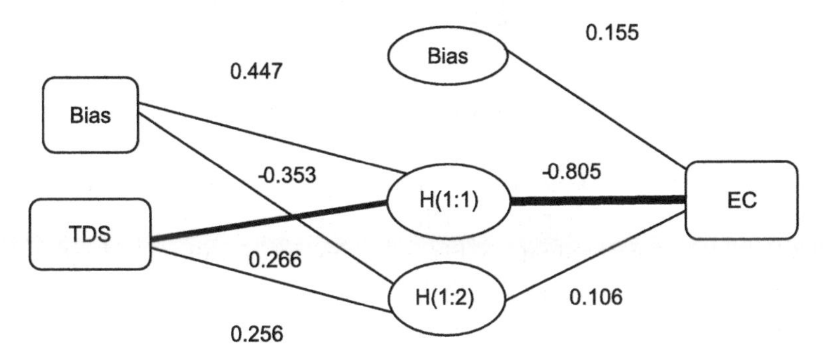

Fig. 3. ANN structure for pre-monsoon season

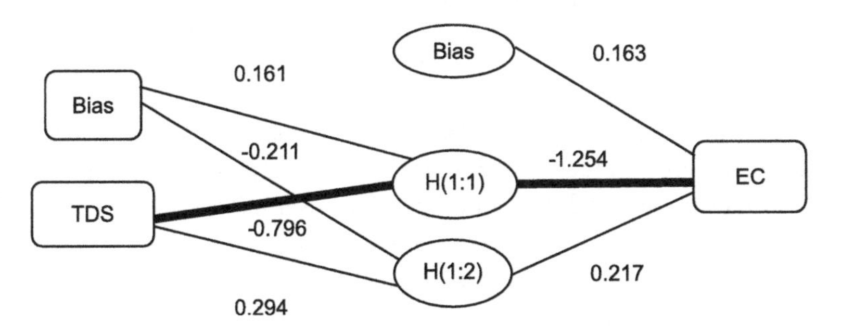

Fig. 4. ANN structure for post-monsoon season

Model Assessment and Evaluation Criteria

The datasets are divided into two datasets i.e. training dataset and testing dataset. For training, 70% of the available data is used. For testing remaining 30% of the available dataset is used. The performances of the developed models are judged based on some evaluation criteria in both training and testing set. Following five performance indices are selected in this study:

Average percent error (APE): APE is the computed average of percentage errors by which forecasts of a model differ from actual values of the quantity being forecast.

$$APE = \left[\frac{\sum_{i=1}^{n} (y_i^{obs} - y_i^{sim})}{n} \right] \tag{1}$$

Nash-Sutcliffe efficiency (NSE): NSE is a normalized statistic that determines the relative magnitude of the residual variance compared to the measured data variance (Nash and Sutcliffe, 1970). NSE indicates how well the plot of observed versus simulated data fits the 1:1 line.

$$NSE = 1 - \left[\frac{\sum_{i=1}^{n} (y_i^{obs} - y_i^{sim})^2}{\sum_{i=1}^{n} (y_i^{obs} - y_i^{mean})^2} \right] \tag{2}$$

Percent bias (PBIAS): PBIAS measures the average tendency of the simulated data to be larger or smaller than their observed counterpart. Positive values indicate model underestimation bias, and negative values indicate model overestimation bias (Gupta et al., 1999).

$$PBIAS = \left[\frac{\sum_{i=1}^{n} (y_i^{obs} - y_i^{sim}) \times (100)}{\sum_{i=1}^{n} y_i^{obs}} \right] \tag{3}$$

Root-mean-square error (RMSE): RMSE is a frequently used measure of the differences between values (sample and population values) predicted by a model or an estimator and the values actually observed.

$$RMSE = \sqrt{\frac{\sum_{i=1}^{n} (y_i^{obs} - y_i^{sim})^2}{n}} \tag{4}$$

The results obtained using regression models in Tables 3 and 4 are presented in Tables 5 and 6 respectively. Results of regression models in Table 4 are presented in Tables 7 and 8 respectively. Fuzzy model result is included for EC prediction using TDS as one input. As a result for fuzzy model is inferior compared to regression models, it is not included in EC predictions using two inputs.

Table 5. Different model evaluation criteria for prediction of EC using TDS only in pre-monsoon season

Regression models	Correlation coefficient (R)		APE (%)		PBIAS (%)		ME_{NASH}		RMSE (%)	
	Train	Test	Train	Test	Train	Test	Train	Test	Train	Test
Linear	0.90	0.20	19.08	44.27	16.14	43.04	0.84	-1.69	110.44	310.34
Exponential	0.90	0.22	27.67	30.19	31.46	31.50	0.29	-0.17	231.97	204.40
Power	0.90	0.22	18.15	43.75	15.53	42.30	0.84	-1.74	108.38	430.20
Polynomial	0.90	0.20	42.60	37.13	47.68	41.11	-0.48	-0.46	334.97	228.95
Fuzzy	0.76	0.13	36.02	62.63	29.87	39.51	0.58	-0.99	175.70	192.78
ANN	0.91	0.24	33.03	34.24	26.77	26.91	0.47	-0.20	200.61	148.99

Table 6. Different model evaluation criteria for prediction of EC using TDS only in post-monsoon season

Regression Models	Correlation coefficient (R)		APE (%)		PBIAS (%)		ME_{NASH}		RMSE (%)	
	Train	Test	Train	Test	Train	Test	Train	Test	Train	Test
Linear	0.76	0.81	19.07	58.22	17.12	34.80	0.58	0.20	121.97	169.19
Exponential	0.74	0.78	24.84	67.84	20.78	43.45	0.51	-0.22	132.62	208.66
Power	0.76	0.81	19.59	58.81	17.34	35.08	0.58	0.19	122.33	169.95
Polynomial	0.76	0.81	56.31	129.39	51.96	97.68	-0.65	-1.67	286.56	437.75
Fuzzy	0.44	0.39	27.61	37.73	24.50	36.79	0.05	-0.08	176.83	267.01
ANN	0.77	0.82	18.13	55.21	16.77	32.28	0.58	0.28	122.11	160.99

CONCLUSION

The spatiotemporal analysis using different kriging method is performed. It was found the results obtained by spherical interpolations are much more precise to actual values in comparison of other kriging interpolations. The maximum values of TDS, EC and Na using different kriging are found at Banat and nearby areas, maximum pH was observed at Makhmulpur, maximum NO_3 at Harhar Fatehpur and Kheda. Considering the importance of conductivity in irrigation prediction model was formed to predict conductivity using TDS alone and with TDS and AMSL. Various regression and fuzzy are used to form prediction model. The ANN was performing better compared to other models in terms of Model efficiency or Nash-Sutcliffe efficiency.

REFERENCES

Ahmadi, S.H. and Sedghamiz, A. (2007). "Geostatistical Analysis of Spatial and Temporal Variations of Groundwater Level". *Environmental Monitoring and Assessment,* 129, pp. 277-294. doi:10.1007/s10661-006-9361-z

Ayazi, M.H., Pirasteh, S., Arvin, A.K.P., Pradhan, B., Nikouravan, B. and Mansor, S. (2010). "Disasters and Risk Reduction in Groundwater: Zagros Mountain Southwest Iran using Geoinformatics Techniques". *Disaster Advances,* 3 (1), pp. 51-57, January.

David, M. (1977). *Geostatistical Ore Reserve Estimation.* Elsevier: Amsterdam.

De Marsily G. (1984). "Spatial Variability of Properties in Porous Media: A Stochastic Approach". In: Bear J. and Corapcioglu M. (eds.) *Fundamentals of Transport Phenomena in Porous Media.* Dordrecht: Martinus Nijhoff, pp. 721-769.

El Afly, M. (2012). "Integrated Geostatistics and GIS Techniques for Assessing Groundwater Contamination in Al Arish area, Sinai, Egypt". *Arabian Journal of Geosciences,* 5(2), pp. 197-215. doi:10.1007/s12517-010-0153-y

Kumar V. and Remadevi. (2006). "Kriging of Groundwater Levels: A Case Study". *Journal of Spatial Hydrology,* 6(1), pp. 81-92.

Machiwal, D., Mishra, A., Jha, M.K., Sharma, A. and Sisodia, S.S. (2012). "Modeling Short-Term Spatial and Temporal Variability of Groundwater Level using Geostatistics and GIS". *Natural Resources Research,* 21(1), pp. 117-136. doi:10.1007/s11053-011-9167-8

Manap, M.A., Nampak, H., Pradhan, B., Lee, S., Sulaiman, W.N.A. and Ramli, M.F. (2012). "Application of Probabilistic-Based Frequency Ratio Model in Groundwater Potential Mapping using Remote Sensing Data and GIS". *Arabian Journal of Geosciences,* 7(2), pp.711-724. doi:10.1007/s12517-012-0795-z

Manap, M.A., Sulaiman, W.N.A., Ramli, M.F., Pradhan, B. and Surip, N. (2013). A Knowledge Driven GIS Modelling Technique for Prediction of Groundwater Potential Zones at the Upper Langat Basin, Malaysia. *Arabian Journal of Geosciences,* 6(5), pp. 1621–1637.

Marko, K., Al-Amri, N.S. and Elfeki, A.M.M. (2014). Geostatistical Analysis Using GIS for Mapping Groundwater Quality: Case Study in the Recharge Area of Wadi Usfan, Western Saudi Arabia. *Arabian Journal of Geosciences,* 7(12), pp. 5239-5252. doi: 10.1007/s12517-013-1156-2

McNeely, R.N., Nelamnis, V.P. and Dwyer, L. (1979). Water Quality Source Book: A Guide to Water Quality Parameter. *Inland Waters Directorates, Water Quality Branch,* Ottawa.

Mehrjardi, R.T., Jahromi, M.Z., Mahmodi, S. and Heidari, A. (2008). "Spatial Distribution of Groundwater Quality With Geostatistics (case study: Yazd-Ardakan Plain). *World Applied Sciences Journal*, 4(1), pp. 9-17.

Mitra, A.K., Nath, S. and Sharma, A.K. (2008). "Fog Forecasting using Rule-based Fuzzy Inference System". *Journal of the Indian Society of Remote Sensing*, 36(3), pp. 243-253.

Nash, J.E. and Sutcliffe, J.V. (1970). River Flow Forecasting through Conceptual Models, part I: A Discussion of Principles. *Journal of Hydrology*, 10(3), pp. 282-290. doi:10.1016/0022-1694(70)90255-6

Nayanaka, V.G.D., Vitharana, W.A.U. and Mapa, R.B. (2010). Geostatistical analysis of soil properties to support spatial sampling in a paddy growing alfisol. *Tropical agricultural research*, 22(1), pp. 34-44. doi:10.4038/tar.v22i12668

Neshat, A., Pradhan, B., Pirasteh, S. and Shafri, H.Z.M. (2013). Estimating Groundwater 805 Vulnerability to Pollution using a Modified DRASTIC Model in the Kerman Agricultural 806 Area, *Iranian Journal of Earth Sciences*, doi: 10.1007/s12665-013-2690-7

Olea, R.A. (1999). *Geostatistics for Engineers and Earth Scientists*. Boston: Kluwer Academic.

Pradhan, B. (2009). Ground Water Potential Zonation for Basaltic Watersheds using Satellite Remote Sensing Data and GIS Techniques. *Central European Journal of Geosciences*, 1(1), pp. 120-129. doi:10.2478/v10085-009-0008-5

Shamsudduha, M. (2007). Spatial Variability and Prediction Modelling of Groundwater Arsenic distributions in the Shallowest Alluvial Aquifers in Bangladesh. *Journal of Spatial Hydrology*, 7(2), pp. 33-46.

Umar, R. (2010). Groundwater Flow Modelling and Aquifer Vulnerability Assessment Studies in Yamuna–Krishna Sub-basin, Muzaffarnagar District. Completion Report Submitted to Indian National Committee on Ground Water Central Ground Water Board (CGWB) Ministry of Water Resources (Govt. of India) (Project No. 23/36/2004-R&D).

Zadeh, L.A. (1965). "Fuzzy Sets". *Information and Control*, 8(3), pp. 338-353. doi:10.1016/S0019-9958(65)90241-X

Use of Soft Computing for Prediction of GWQI: A Case Study Using Artificial Neural Network

Nikunj Ashiyani[1] and T.M.V. Suryanarayana[2]

ABSTRACT

Groundwater is a major source of water in Matar Taluka for irrigation and water supply. The groundwater quality largely depends on the parameters like pH, TDS, chlorine etc. There are 778 groundwater samples are analysed to calculate the groundwater quality index value. This paper presents Artificial Neural Network (ANN) Models that might be used to predict groundwater quality index value. These models were trained using the seven different training algorithms, namely Cascade Forward Backpropagation, Elman Backpropagation, Feed Forward Backpropagation, Feed Forward Distributed Time Delay, Generalized Regression, Layer Recurrent, NARX. This study shows that ANN can be used to forecast groundwater quality index accurately. As per the classification based on groundwater quality index 26.09 per cent ground water samples shows good quality of water and 9.90 per cent samples shows excellent quality of ground water. The ANN models with Feed Forward Distributed Time Delay training algorithm yields RMSE and r value of 1.675 and 0.999 respectively during training and RMSE and r value of 0.417 and 1 during validation.

Keywords: *Groundwater quality index, GWQI, matar, artificial neural network, feed forward distributed time delay*

INTRODUCTION

Water resources have been the most exploited natural systems, since man strode the earth. As a result of increasing, civilization, urbanization, industrialization and other developmental activities, our natural water system is being polluted by different sources. The pollutants coming as a waste to the water bodies are

1. P.G. Student, Water Resources Engineering and Management Institute, M.S. University of Baroda, Samiala – 391 410, Ta. & Dist.: Vadodara.
2. Associate Professor, Water Resources Engineering and Management Institute, M.S. University of Baroda, Samiala – 391 410, Ta. & Dist.: Vadodara.

likely to create nuisance by way of physical appearance, odour, taste, quality and render the water harmful for utility. This has resulted in the decrease in the quality of drinking water available. The ground water quality is normally characterized by different physical and chemical characteristics. These parameters change widely due to the type of pollution, seasonal fluctuation, ground water extraction, etc. Monitoring of water quality levels is thus important to assess the levels of pollution and also to assess the potential risk to the environment.

Groundwater quality index is one of the most effective tools to communicate information on the quality of water to the concerned citizens and policy makers. It, thus, becomes an important parameter for the assessment and management of groundwater. Artificial Neural Network (ANN) is an example of information processing structures that have been conceived in the field of neurocomputing. Neurocomputing is the technological discipline concerned with information processing systems that autonomously develop operational capabilities in adaptive response to an information environment. Neurocomputing is also known as parallel distributed processing. Figure 1 shows a simple artificial neural network.

The basic building block of an artificial neural network is an artificial neuron. As shown in Fig. 2, a typical artificial neuron consists of three components: input links (W_1, W_2,..... W_n), a processing unit (the circle) and an output link (W_0). The processing unit is divided into two components. The side connected to the input links has a summation function and a threshold (W_t), which is a negative resident weight of the neuron. The summation function sums up the inputs and the threshold, and feeds the sum into the other side of neuron that has an activation function to produce an output stimulus. This output stimulus is then communicated to outside through the output link.

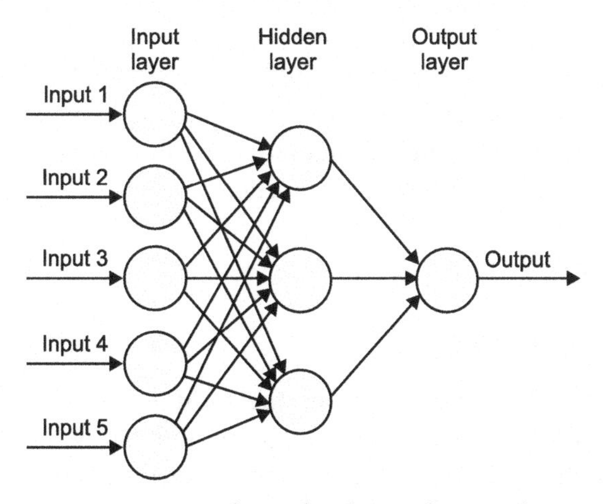

Fig. 1. A simple artificial neural network

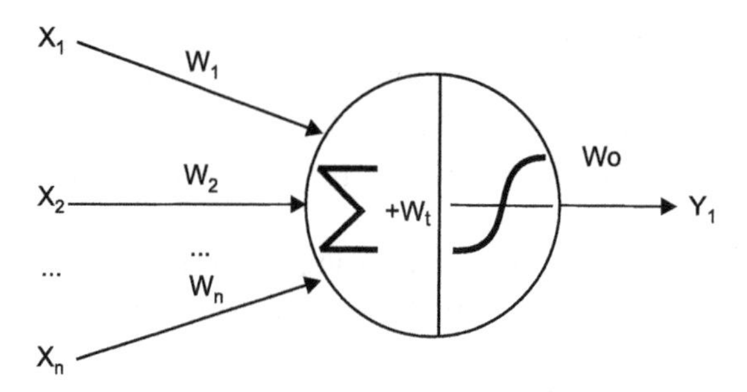

Fig. 2. A typical artificial neuron

Abdul Hannan et al. (2010) used the two types of Artificial Neural Networks (ANNs), Generalized Regression Neural Network (GRNN) and Radial Basis Function (RBF) have been used for heart disease to prescribe the medicine. Goyal and Goyal (2012) developed Radial basis (exact fit) artificial neural network model for estimating the shelf life of burfi stored at 30°C. Kisi and Ay (2012) found the RBNN model to be better than the ANFIS model in estimation of DO concentration. Bessaih et al. (2014) developed the Artificial network models to forecast monthly water levels for three wells in Wadi El Jezzy Catchment. Ashiyani and Suryanarayana (2015) studied and presented the results on assessment of Groundwater Quality using Groundwater Quality Index (GWQI) method.

STUDY AREA

Gujarat state is located in the western part of India. In this study, Matar Taluka of Kheda district area is selected. The Kheda district is located (between 72° 32' to 73° 37' East longitude and between 22°30' to 23° 18' North latitude) in Gujarat. The study was conducted in a pre-monsoon season and post-monsoon season of ten years from 1997 to 2006. Total 778 samples from different localities of Matar Taluka were collected. Seven groundwater quality parameters are selected i.e. pH, calcium, magnesium, chloride, sulphate, total dissolved solids and sodium absorption ratio.

METHODOLOGY

Groundwater Quality Index

Groundwater quality index is one of the most effective tools to monitor the surface as well as ground water pollution and can be used efficiently in the implementation of water quality upgrading programmes. It is one of the aggregate indices that have been accepted as a rating that reflects the composite influence on the overall quality of numbers of precise water quality characteristics.

For computing GWQI three steps are followed. In the first step, each of the parameters has been assigned a weight (w_i) according to its relative importance in the overall quality of water for drinking purposes (Table 1). The maximum weight of 4 has been assigned to the parameters like TDS, pH and sulphate due to its major importance in water quality assessment. SAR and magnesium which is given the minimum weight of 2 as magnesium by itself may not be harmful. In the second step, the relative weight (W_i) is computed using the following equation:

$$W_i = \frac{w_i}{\Sigma_{i=1}^{n} w_i}$$

Where,

W_i is the relative weight, w_i is the weight of each parameter and n is the number of parameters. Calculated relative weight (W_i) values of each parameter are also given in Table 1.

Table 1: Relative weight (W_i) values of each parameter

Parameters	Indian standards	Weight (w_i)	Relative weight (W_i)
pH	6.5-8.5	4	0.19047619
Chloride	250-1000	3	0.142857143
Sulphate	200-400	4	0.19047619
Ca^{+2}	75-200	2	0.095238095
Mg^{2+}	30-100	2	0.095238095
SAR	10-255	2	0.095238095
TDS	500-2000	4	0.19047619
Total		21	1

In the third step, a quality rating scale (q_i) for each parameter is assigned by dividing its concentration in each water sample by its respective standard according to the guidelines laid down in the BIS and the result multiplied by 100.

$$q_i = \left(\frac{C_i}{S_i}\right) \times 100 \qquad (1)$$

Where,

q_i is the quality rating, C_i is the concentration of each chemical parameter in each water sample in mg/L, S_i is the Indian drinking water standard for each chemical parameter in mg/L according to the guidelines of the BIS.

For computing the WQI, the SI is first determined for each chemical parameter, which is then used to determine the WQI as per the following equations.

$$SI_i = W_i \times q_i \tag{2}$$

$$WQI = \sum_{i=1}^{n} SI_i \tag{3}$$

Where,

SI_i is the sub-index of ith parameter, q_i is the rating based on concentration of ith parameter, n is the number of parameters.

The computed WQI values are classified into five types, "excellent water" to "unsuitable for drinking".

Table 2. Water quality classification based on WQI value

WQI value	Water quality
<50	Excellent water
50–100	Good water
100–200	Poor water
200–300	Very poor water
>300	Unsuitable for drinking

ARTIFICIAL NEURAL NETWORK (ANN)

Neural network inputs typically range from 0 to 1 (inclusive) and usually the output ranges from 0 to 1. Hence, the input and output vector values are converted in the range of 0 to 1 using the following equation for data normalization and data denormalization.

$$X_n = \frac{X - X_{min}}{X_{max} - X_{min}}$$

Where,

X_n = Normalized input (or output) value

X = Actual input (or output) value

X_{max} = Maximum value of input (or output)

X_{min} = Minimum value of input (or output)

Development of groundwater quality index prediction model using artificial neural network.

Three models are developed to predict Groundwater Quality index of Matar Taluka using ANN.

ANN Model-1: 60%-40%

ANN Model-2: 70%-30%

ANN Model-3: 80%-20%

The models developed using ANN for Matar Taluka with training and validations are described in detail below,

Groundwater quality index prediction model is developed in the following manner. Import input and target data either from workspace or from file to ANN data manager. Network is created using ANN tool using different training algorithms, i.e. cascade forward backpropagation, Elman backpropagation, feed forward backpropagation, feed forward distributed time delay, generalized regression, layer recurrent, linear layer (design), NARX and radial basis (Exact Fit).

Cascade Forward Back propagation Network

Cascade forward back propagation model shown in Fig. 3 is similar to feed-forward networks, but include a weight connection from the input to each layer and from each layer to the successive layers. While two-layer feed forward networks can potentially learn virtually any input-output relationship, feed-forward networks with more layers might learn complex relationships more quickly. Cascade forward back propagation ANN model is similar to feed forward back propagation neural network in using the back propagation algorithm for weights updating, but the main symptom of this network is that each layer of neurons related to all previous layer of neurons.

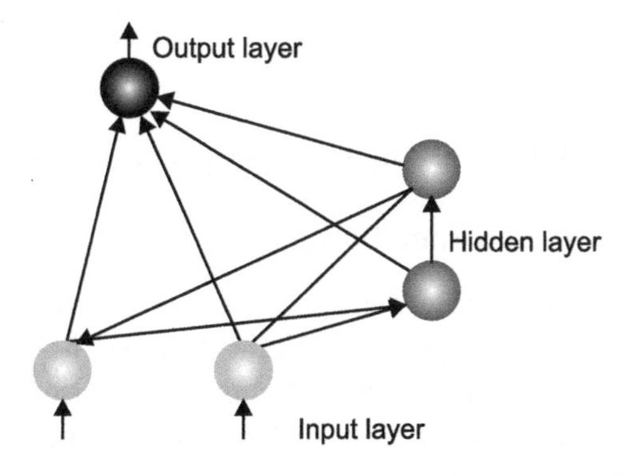

Fig. 3. Cascade forward back propagation network

Elman Back Propagation

Elman networks are feed-forward networks with the addition of layer recurrent connections with tap delays. Elman networks with one or more hidden layers can learn any dynamic input-output relationship arbitrarily well, given enough neurons in the hidden layers. However, Elman networks use simplified derivative calculations at the expense of less reliablelearning.

Feed Forward Back Propagation

Feed forward back propagation artificial neural network model shown in Fig. 4 consists of input, hidden and output layers. Back propagation learning algorithm was used for learning these networks. During training this network, calculations were carried out from input layer of network toward output layer, and error values were then propagated to prior layers. Feed forward networks often have one or more hidden layers of sigmoid neurons followed by an output layer of linear neurons. Multiple layers of neurons with nonlinear transfer functions allow the network to learn nonlinear and linear relationships between input and output vectors. The outputs of a network such as between 0 and 1 are produced, then the output layer should use a sigmoid transfer function (such as logsig).

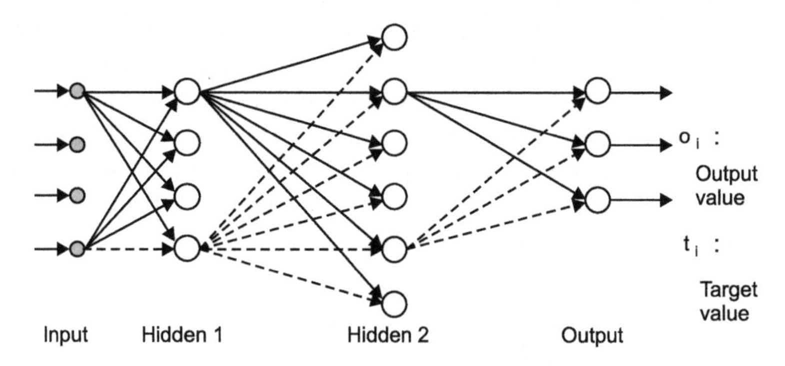

Fig. 4. Feed forward back propagation network

Feed Forward Distributed Time Delay

Feed forward distributed time delay networks are similar to feed forward networks, except that the input weight has a tap delay line associated with it. This allows the network to have a finite dynamic response to time series input data. This network is also similar to the distributed delay neural network, which has delays on the layer weights in addition to the input weight.

Generalized Regression

A GRNN is a variation of the radial basis neural networks, which is based on kernel regression networks. A GRNN does not require an iterative training procedure as back propagation networks. It approximates any arbitrary function between input and output vectors, drawing the function estimate directly from the training data. In addition, it is consistent that as the training set size becomes large, the estimation error approaches zero, with only mild restrictions on the function. A GRNN consists of four layers: input layer, pattern layer, summation layer and output layer as shown in Fig. 5. The number of input units in input layer depends on the total number of the observation parameters. The first layer is connected to the pattern layer and in this layer each neuron presents a training pattern and its output. The pattern layer is connected to the summation layer. The summation layer has two different types of summation, which are a single division unit and summation units. The summation and output layer together perform a normalization of output set.

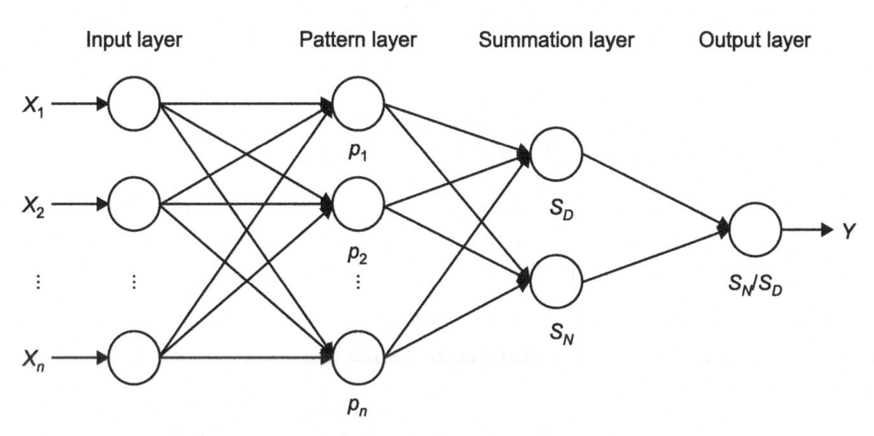

Fig. 5. General structure of GRNN

Layer Recurrent

Layer recurrent neural networks are similar to feed forward networks, except that each layer has a recurrent connection with a tap delay associated with it. This allows the network to have an infinite dynamic response to time series input data. This network is similar to the time delay and distributed delay neural networks, which have finite input responses.

NARX

NARX (Nonlinear autoregressive with external input) networks can learn to predict one time series given past values of the same time series, the feedback input, and another time series, called the external or exogenous time series.

RESULTS AND ANALYSIS

Table 3. Water quality classification based on GWQI value

GWQI value	Water quality	Percentage of water samples		
		Pre-monsoon	Post-monsoon	Overall
<50	Excellent water	9.794	10	9.90
50-100	Good water	24.485	27.692	26.09
100-200	Poor water	31.186	34.615	32.90
200-300	Very poor water	23.969	20.513	22.24
>300	Unsuitable for drinking	10.567	7.179	8.87

As shown in Table 3, the GWQI for 778 ground water samples ranges from 29.05 to 640.75, with almost 64.01 percent of the samples exceeded 100, the upper limit for drinking water. About 32.90 per cent of water samples are poor in quality and 22.24 per cent of water samples are of very poor quality and should not use directly for drinking purpose. As per the classification based on groundwater quality index 26.09 per cent ground water samples shows good quality of water and 9.90 per cent samples shows excellent quality of ground water.

ARTIFICIAL NEURAL NETWORK (ANN)

After developing the best Groundwater quality index prediction Models using ANN for Matar Taluka with three different combinations for training and validation, the results are obtained in Table 4, Table 5 and Table 6.

Table 4. RMSE and r values for different training algorithms of ANN Model-1 for Matar Taluka

Network	Training (60%)		Validation (40%)	
	RMSE	r	RMSE	r
Cascade forward back propagation	4.9703	0.9989	5.974956	0.997933
Elman back propagation	13.214	0.9924	1.697972	0.999848
Feed forward back propagation	28.165	0.9647	136.4199	0.164129
Feed forward distributed time delay	0.5492	1	1.232054	0.999911
Generalized regression	104.73	0.9759	88.88885	0.998142
Layer recurrent	173.5961	$-5.3E^{-16}$	3.015451	0.999481
NARX	5.5775	0.9987	6.730629	0.997564

Table 5. RMSE and r value for different training algorithms of ANN Model-2 for Matar Taluka

Network	Training (70%)		Validation (30%)	
	RMSE	r	RMSE	r
Cascade forward back propagation	8.77	1	3.953	0.999
Elman back propagation	3.04	1	0.594	0.999
Feed forward back propagation	14.366	0.991	2.635	0.999
Feed forward distributed time delay	2.47	1	2.008	0.999
Generalized regression	104.91	0.979	84.279	0.998
Layer recurrent	9.427	0.996	4.452	0.999
NARX	12.518	0.994	3.829	0.999

Table 6. RMSE and r value for different training algorithms of ANN Model-3 for Matar Taluka

Network	Training (80%)		Validation (20%)	
	RMSE	r	RMSE	r
Cascade forward back propagation	9.9399	0.9957	3.6579	0.999
Elman back propagation	170.5648	0.167546	1.9533	0.9997
Feed forward back propagation	15.01898	0.990093	1.024277	0.99992
Feed forward distributed time delay	1.674762	0.999879	0.4175	1
Generalized regression	104.56	0.9812	74.76	0.9974
Layer recurrent	3.85038	0.999353	1.1619	0.9999
NARX	171.703	-3.8E-16	0.8669	0.9999

The comparison is carried out to conclude the best ANN model among all developed models for particular station and is given in Table 7.

Table 7. Best ANN model developed for Matar Taluka

Taluka	Phase	ANN Model-1		ANN Model-2		ANN Model-3	
		RMSE	r	RMSE	r	RMSE	r
Matar	Training	0.549	1	2.47	1	1.675	0.99
	Validation	1.232	0.999	2.008	0.99976	0.417	1

The above Table shows that the ANN Models with feed forward distributed time delay training algorithm having RMSE and r value of 1.675 and 0.999 respectively during training and RMSE and r value of 0.417 and 1 during validation. Hence, ANN Model-3 is the best model for Matar Taluka for Groundwater quality index prediction.

Charts given below shows correlation between predicted GWQI vs. observed GWQI of ANN Model-3 during training and validation for Matar Taluka. It is observed that during training and validation, predicted GWQI and observed GWQI values are found to be highly correlated.

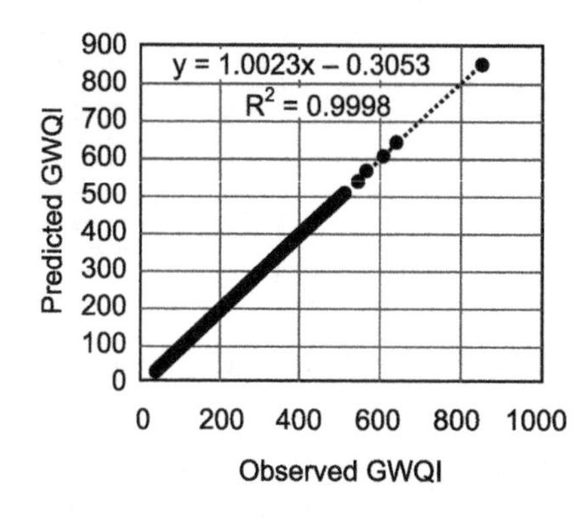

Correlation between Predicted GWQI vs. Observed GWQI of ANN Model-3 for Matar Taluka during training

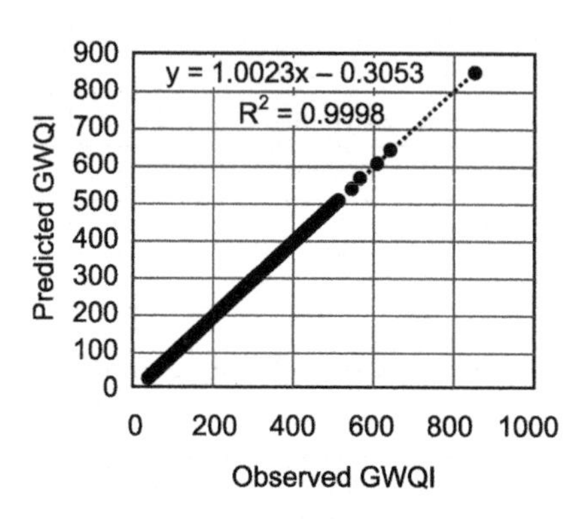

Correlation between Predicted GWQI vs. Observed GWQI of ANN Model-3 for Matar Taluka during validation

CONCLUSION

As per the classification based on groundwater quality index 26.09 per cent ground water samples shows good quality of water and 9.90 percent samples shows excellent quality of ground water. The ANN Models with feed forward distributed time delay training algorithm yields RMSE and r value of 1.675 and 0.999 respectively during training and RMSE and r value of 0.417 and 1 during validation. The same may be used for groundwater quality index prediction in Matar Taluka.

REFERENCES

Ashiyani, Nikunj and Suryanarayana, T.M.V. (2015). "Assessment of Groundwater Quality using GWQI Method", *Proceedings of National Conference on "Recent Research & Development in Core Disciplines of Engineering"*, Vadodara Institute of Engineering, March 2015.

Bessaih, Nabil, Qureshi, Mohsin, Al-Jabri, Fatima Salem, Al-Harmali, Iman Rashid and Al Naamani. Zahra Ali. (2014). "Groundwater Water Level Prediction in Wadi El Jezzy Catchment Using ANN", *Proceedings of the World Congress on Engineering*, Vol. 1.

Goyal, Sumit and Goyal, Gyanendra Kumar. (2012). "Radial Basis (Exact Fit) Artificial Neural Network Technique for Estimating Shelf Life of Burfi", *Advances in Computer Science and its Applications,* 1(2), pp. 93-96, June.

Ozgur, Kisi and Murat, A.Y. (2012). "Comparison of ANN and ANFIS Techniques in Modeling Dissolved Oxygen", *Sixteenth International Water Technology Conference, IWTC-16,* Istanbul, Turkey.

Shaikh, Abdul Hannan, Manza, R.R. and Ramteke, R.J. (2010). "Generalized Regression Neural Network and Radial Basis Function for Heart Disease Diagnosis", *International Journal of Computer Applications,* 7(13), October.

Application of Artificial Neural Network (ANN) Technique in Transportation Engineering

Gaurang J. Joshi[1]

INTRODUCTION

All the real world situations are characterized by complexity through the non-linear behaviour. Moreover, the relationships between most of the controlled as well as uncontrolled variables are dynamic in nature and have varying degree of uncertainty either due to randomness or vagueness. Much of the decision making in the real world takes place in an environment in which goals, the constraints and the consequences of possible actions are unknown precisely. Major efforts in engineering decision making involves understanding the behaviour of a decision variable depending on the variables influencing the decision variable. This understanding is useful in predicting the likely changes in future due to changes in the system characteristics. Transportation engineering also deals with the situations when the relationships between variables are not strictly governed by any law of physics; particularly for travel demand estimation as well as travel time prediction. Both of these situations are significantly affected by human decisions and societal values and hence becomes difficult to model with conventional cause-effect mathematical modelling approach laden with number assumptions about the variables' nature and their relationship. Under the circumstances, technique mimicking functioning of human brain comes to the rescue of modelers and can be applied effectively for prediction of travel demand and travel time on road corridors. The present article tries to highlight basic feature of artificial neural network and shows its usefulness by two case studies related to travel demand estimation by trip generation modelling and travel time for an urban roadway.

BACKGROUND OF ANN

An artificial neuron is a computational model inspired in the natural neurons. Natural neurons receive signals through synapses located on the dendrites or membrane of the neuron. When the signals received are strong enough (surpass

1. Associate Professor, Transportation Engineering & Planning Civil Engineering Department, SVNIT, Surat.

a certain threshold), the neuron is activated and emits a signal through the axon. This signal might be sent to another synapse, and might activate other neurons.

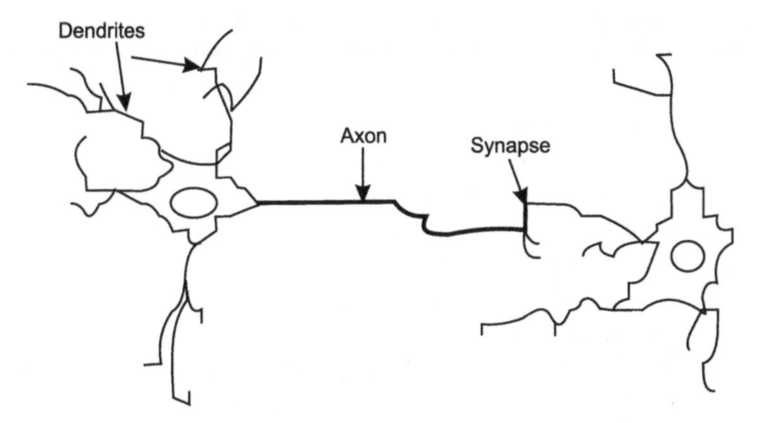

Fig. 1. Natural neurons (artist's conception)

The complexity of real neurons is highly abstracted when modelling artificial neurons. These basically consist of inputs (like synapses), which are multiplied by weights (strength of the respective signals), and then computed by a mathematical function which determines the activation of the neuron. Another function (which may be the identity) computes the output of the artificial neuron (sometimes in dependence of a certain threshold). ANNs combine artificial neurons in order to process information. Natural neurons like axon, dendrites and synapse are clearly depicted in figure 1.

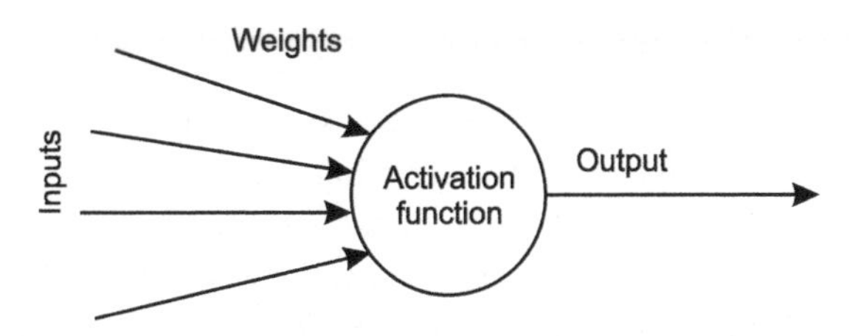

Fig. 2. An artificial neuron

The higher the weight of an artificial neuron is, the stronger the input which is multiplied by it will be. The overall functioning of an artificial neuron in described in figure 2. Weights can also be negative, so we can say that the signal is inhibited by the negative weight. Depending on the weights, the computation of the neuron will be different. By adjusting the weights of an artificial neuron we can obtain the output we want for specific inputs. But when we have an ANN of hundreds or thousands of neurons, it would be quite complicated to find by hand all the necessary weights. But we can find algorithms which can adjust the weights of the ANN in order to obtain the desired output from the network. This process of adjusting the weights is called learning or training. Learning is a process by which the free parameters of a neural network are adapted through a continuing process of stimulation by the environment in which the network is embedded. The type of learning is determined by the manner in which the parameter changes take place.

During the process of learning, the network adjusts its parameter, the synaptic weights in response to an input stimulus so that its actual output response converges to the desired output response. When the actual output response is the same as the desired one, the network has completed the learning phase and has acquired knowledge. As different learning methodologies suit different people, so do different learning techniques suit different artificial neural networks. The classification of the learning process in neural network is shown in figure 3.

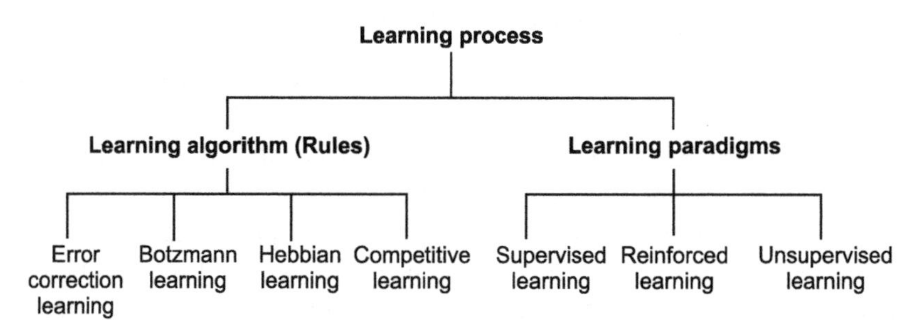

Fig. 3. Classification of learning process

The purpose of the activation function is to ensure that the neurons response is bounded, that is, the actual response of the neuron is conditioned, or damped, as a result of large or small activating stimuli and thus is controllable. The various types of activation function used are shown in Figure 4.

The two most popular activation functions are hard limiter and the sigmoid. All the functions depicted in figure 4 are all bounded they have an upper and/or lower limit such as ±1, ±2. In actual networks the users choose the value of the bounds.

FEATURES OF ANN

ANN's have several attractive features:

- Their ability represent non-linear relations make them well suited for non-linear modelling in control systems.

Linerar (ramp) function	$f(x) = \alpha x$ Where α is constant. 1. This function has linear zone. 2. This function is easy to implement.
Non-linear (ramp) function	$f(x) = 1$ if $x \geq 1$ $= x$ if $\lvert x \rvert < 1$ $= -1$ if $x \leq 1$ This function is used to represent the simplified non-linear operation.
Step function (hard limiting)	$f(x) = +1$ if $x > 0$ $= -1$ otherwise 1. This function is fast and easy to implement 2. It has linear range and cannot smoothly imitate functions.
Sigmoid (S-shaped) function	1. This function is continuously differentiable. 2. This function can make fuzzy decisions. 3. This function is not easy to implement. 4. The output of this function is limited to positive values.
Hyperbolic tangent function	$f(x)$ tanh(x) $f(x)$ x

Fig. 4. Typical activation functions

- Adaptation and learning in uncertain system through offline and online weight adaptation.
- Parallel processing architecture allows fast processing for large scale dynamic systems.
- Neural network can handle large number of inputs and can have many outputs.
- ANN can store knowledge in a distributed fashion and consequently have a high fault tolerance.

APPLICATIONS OF ANN

Character Recognition—The idea of character recognition has become very important as handheld devices like the Palm Pilot are becoming increasingly popular. Neural networks can be used to recognize handwritten characters.

Image Compression—Neural networks can receive and process vast amounts of information at once, making them useful in image compression. With the Internet explosion and more sites using more images on their sites, using neural networks for image compression is worth a look.

Stock Market Prediction—The day-to-day business of the stock market is extremely complicated. Many factors weigh in whether a given stock will go up or down on any given day. Since neural networks can examine a lot of information quickly and sort it all out, they can be used to predict stock prices.

Travelling Salesman's Problem—Interestingly enough, neural networks can solve the travelling salesman problem, but only to a certain degree of approximation.

Medicine, Electronic Nose, Security and Loan Applications—These are some applications that are in their proof-of-concept stage, with the acceptance of a neural network that will decide whether or not to grant a loan, something that has already been used more successfully than many humans.

Miscellaneous Applications—These are some very interesting (albeit at times a little absurd) applications of neural networks.

APPLICATION OF ANN TECHNIQUE IN TRANSPORTATION ENGINEERING

Artificial Neural Network (ANN) systems have been widely applied in various fields of transportation engineering such as travel behaviour, traffic flow study and management. Artificial neural networks are employed for modelling the relationship that exist among driver injury severity and crash causes or factors that have to do with the driver, vehicle, roadway and the environment characteristics. The use of artificial neural networks can reveal the relationship that exists between vehicle, roadway and environment characteristics and driver injury severity. Traffic forecasting problems involving complex interrelationships between variables of traffic system can be efficiently solved using ANN. They provide realistic and fast ways for developing models with enough data. The ANN models help us to compare the states' road safety performance by the number of motor vehicle fatalities. Used in many fields, the application of ANNs has seen a lot of success in a number of different areas of specialization, including transportation engineering. Researches have been done to find the relationship that exists between driver injury severity and driver, vehicle, road and environment characteristics, using two well-known neural network paradigms, the multilayer perceptron and the fuzzy adaptive resonance theory neural networks. Recently, ANN have been adopted for sequential forecasting of incident duration from the point of view of incident notification to the incident road clearance. Prediction of the lane-change occurrence with respect to freeway crashes using the traffic surveillance data collected from a pair of dual loop detectors, and a study understanding the circumstances under which drivers and passengers are more likely to be killed or more severely injured in an automobile accident, can help to improve the overall driving safety situation. The earliest applications of ANNs in pavement systems concentrated on areas such as planning, traffic control and operations, construction and maintenance and facilities management. The last few years have seen considerable interest in using ANNs for pavement systems analysis— structural and performance prediction— and design.

Before any predictive distress models can be developed, quantitative measures of pavement distress and performance must be established and methods developed to measure their values over time. This is an area in which neural networks have already shown promise. Another area in which ANNs have already been used is pavement classification. Most state highway agencies maintain permanent traffic recorder stations at strategic locations in order to develop a database of traffic patterns for different road types. This database captures the seasonal variations in the monthly average daily traffic (MADT) at each recorder location. In principle, if you can match a road segment where there is no recorder to one stored in the database, you can forecast its average annual daily traffic (AADT) from a short-term traffic count by applying the seasonal variations stored in the database. In practice, the database is condensed into a handful of road types exhibiting similar traffic patterns and road attributes; this makes it easier to find a match in the database. The task of condensing the database is a classic pattern-classification application at which ANNs excel.

Some state highway agencies, such as those in Illinois and North Carolina, have begun monitoring selected in-service pavements for performance. They are keeping records of pavement materials and cross sections, applied traffic loads, and climatic conditions. Similar data have been generated, in even greater volume, from the Long-Term Pavement Performance (LTPP) project of the Strategic Highway Research Program (SHRP). The structural response of the pavements to the recorded loads can be calculated using mechanistic analysis programmes. By using ANNs, engineers can then correlate the observed pavement performance with the calculated structural response. Because ANNs excel at mapping in higher-order spaces, such models can go beyond the existing univariate relationships (such as those based on asphalt flexural strain or subgrade vertical compressive strain). ANNs could be used to examine several variables at once and the interrelationships between them. ANNs could also be used to develop models for distress phenomena such as thermal cracking, block cracking, and rutting in AC pavements, and faulting and D-cracking in concrete pavements.

Predictive pavement distress models, whether they are developed using ANNs or conventional modelling techniques, will have to be calibrated to local conditions. This is done using shift factors, which adjust the predicted distress development to more realistically reflect field-observed pavement distress and performance. These shift factors not only vary from state to state, but will have to be periodically updated for temporal changes in climate, materials, construction specifications, and traffic. Radial basis function networks would be particularly well suited to this task because they can be incrementally retrained. This is an important point. Some ANNs, such as backpropagation networks, must be completely retrained if additional data become available. Others, such as radial basis function and counter propagation networks, can evolve over time to accommodate both new data and changed data. The choice of a network type should anticipate future enhancements as well as current needs. The final area of infrastructural analysis in which ANNs could be used is the modelling of the traffic loads applied to the pavement. Unlike current empirical design procedures based on equivalent axle loads or equivalent aircraft, mechanistic design procedures make it possible to explicitly model the landing gear geometry and wheel loads of each individual vehicle. This requires the user to develop a realistic vehicle mix and project it over the design life of the vehicle. In that capacity, ANNs developed for time-series analysis (such as recurrent neural networks) can play a major role.

CASE STUDY 1: TRAVEL DEMAND ESTIMATION THROUGH ANN

Surat city, the study area has been one of the fastest growing cities in Gujarat. The population changes that the city has undergone the last few decades are phenomenal. Under such circumstances the need for greater number of houses in every size and form has increased. The city spread over 112.28 sq.km with population of 24.33 lakh to date needs division in smaller units to reflect the peculiar, socio-economic and land use growth prevailing. For this, Surat is considered to be bounded by an external cordon as per the boundary of Surat Municipal Corporation (SMC). The election ward is taken as micro level unit for study. Then the zones are built from 2 to 3 ward boundaries as the meso level unit.

The zone boundary delineates through the major road of the city. About 5 to 9 zones from a sector, matching with the administrative zones of the SMC at macro level. The total number of wards, zones and sectors in the study area respectively are 66, 32 and 6 thus 32 study zones are made to cover all the 66-census wards of the city. The Surat city is divided into six sectors: Central, North, East, South, South-West and West.

DEVELOPMENT OF TRIP GENERATION MODEL: Di Tri GeM

Disaggregate Approach

A transportation-planning study cannot possibly trace the patterns of every individual residing within a region. As a result, the geographical patterns of trip making are summarized by dividing the region into smaller travel-analysis zones and by associating the estimated trips with these zones. Early models of trip generation considered the zone to be the smallest entity of interest as far as trip making was concerned. Consequently, these models were calibrated on a zonal basis, meaning that the overall zonal characteristics were used as independent or explanatory variables. These zonal attributes included variables such as the zonal population, the average zonal income, the average vehicle ownership, and average family size, etc. However, this tends to mask intra zonal variability and affects the accuracy of the trip generation estimates.

Further disaggregation to household level in various zones help to capture intra zonal variation in trip generation rate. Therefore in the present study zone wise household data has been used to develop the per capita trip rate model. Disaggregation of dependent variable of daily person trip rate (Y) is made in terms of work purpose trip rate (Y1), education purpose trip rate (Y2) and other purpose (recreation, shopping, social, etc.) trip rate (Y3) in the present study.

Di Tri GeM FORM

The artificial neural network based **Di**saggregate **Tr**ip **Ge**neration **M**odel (DiTriGeM) is developed considering the following set of input and output variables:

Input variables

X_1 = Family size

X_2 = Working members in a family

X_3 = No. of school/college going members in a family

X_4 = Monthly family income in thousand Rs.

X_5 = Vehicle ownership

Output variables

Y_1 = Work trip rate (tpcd)

Y_2 = Education trip rate (tpcd)

Y_3 = Other trip rate (tpcd)

The above model variables are presented in Figure 5 as input and output nodes of neural network. Feed Forward Network (FNN) with error back propagation training algorithm is adopted to develop the neural network model.

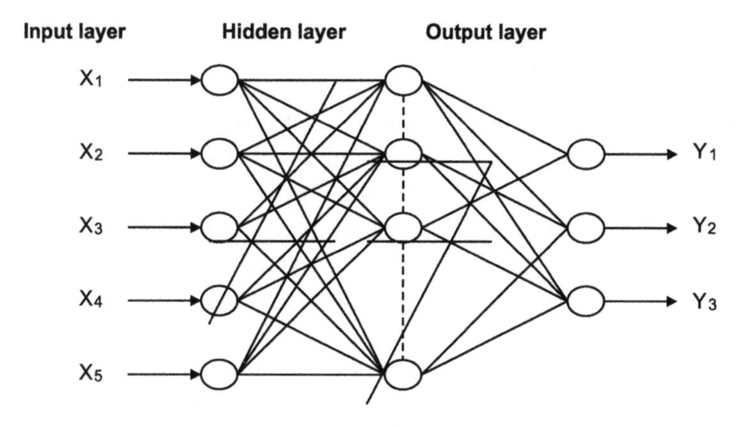

Fig. 5. Di Tri GeM form

MODEL CALIBRATION APPROACH

The calibration data for the model is containing 836 number of travel behaviour survey samples collected in the study area. Out of this 72% i.e. 600 samples are used as training data set and remaining 28% i.e. 236 samples are used as testing data set. These data sets are enclosed as Appendix-D. Stage wise calibration process is as follows:

Stage 1: Normalization

The calibration data are first normalized by applying following formula:

$$(x_i)_{NOR} = \frac{0.10 + (x_i)_{act}}{1.2(x_{max})}$$

Where, $x_{i_{NOR}}$ = Normalize variable

x_{act} = Actual value of variable

$x_{i\,max}$ = Maximum actual value variable

Stage 2: Network Architecture

The number of hidden layers and number of neurons in each hidden layer are to be specified along with the transfer function for each layer. The functions for training of network and learning are also specified. The error goal in terms of mean squared error (MSE) as well as maximum number of training epochs are specified. Various combinations of these network architecture parameters are trained to arrive at the satisfactory network configuration.

Stage 3: Network Training

The normalized training data sets of input matrix (600×5) and target matrix (600×3) are presented to the network. The initialization of weights and bias takes place as per the in-built function. The modification in weight and bias values occurs during the training process till either the target error i.e. MSE, is achieved or maximum number of epochs are reached.

Stage 4: Network Testing

The trained network is presented the testing input data set (236×5) for simulating the observed output.

Stage 5: Performance Measures

Indices of performance measures of trained network are calculated to check the model capability to simulate the desired outputs. The indices like Absolute Relative Error (ARE), Mean Absolute Relative Error (MARE) and Co-relation co-efficient for simulated and targeted output are calculated.

Stage 6: Reproducibility Simulation

The trained network is presented the training input data set (600×5) and output is obtained (600×3). The performance indices for the observed and simulated output are worked out. These measures indicate the ability of network to reproduce the output for the inputs known to the network.

Stage 7: Denormalization of Testing and Training Data

The testing and training data are denormalized by $x_{iact} = 1.2(x_{max})[(x_i)_{NOR} - 0.10]$ and checked for performance indices for training and testing data. When results are acceptable then next step is executed other wise the network architecture is changed (stage: 2) and again the same procedure is carried out.

Stage 8: Freezing the Architecture

When all the performance measures are acceptable, final network architecture in terms of link weights and neuron bias and trains for function is frozen.

Di Tri GeM CONFIGURATION

Network Architecture

Several architectures for calibration were tested. The network with following architecture has been accepted as the Disaggregate Trip Generation Model (DiTriGeM):

Table 1. Network architecture of DiTriGeM

Description	Node	Neurons	Transfer function	Training function	Learning function	MSE	Epoch
Input layer	5	-	-	Trainlm	Trainlm	0.00097	13
First hidden layer	-	11	Logsig				
Second hidden layer	-	15	Logsig				
Output layer	3	-	Purelin				

The architecture of the model is finally written as 5-11-15-3. The mean squared error (MSE) is calculated as,

$$MSE = \sqrt{\frac{\left[\sum_{i=1}^{N}\left(Q^{i}{}_{obs} - \overline{Q}_{pre}\right)^{2}\right]}{N}}$$

Where,

$Q^{i}{}_{obs}$ = Observed input variable

$\overline{Q}_{pre}$ = Mean predicted value of input variable

N = Number of sample

Weights and Bias

The inter neuron link weights and at neurons bias for calibrated ANN model are presented respectively in matrix and vector form. Table 2 contains weights of inter connecting links between input nodes and first hidden layer. The links between first hidden layer and second hidden layer are given as weight matrix in Table 3, where as Table 4 contains weight matrix for connections between second hidden layer and output layer. The bias generated at each node of hidden layers and output layer are presented in form of vectors in Table 5.

The Hinton Graphs showing inter layer connection weights and bias at neurons of a layer are presented in Figures 6, 7 and 8 respectively for first hidden layer and input layer (W1 and B1), for second and first hidden layer (W2 and B2) and output layer and second layers (W3 and B3). The black band in the Hinton graph is showing bias at the nodes where as dark grey band shows inter layer neuron weights. The positive weights and bias are shown by white colour rectangles and negative weights and bias are shown by grey colour rectangles with their sizes proportional to their values.

Table 2. Weight matrix (W1): Input layer to first hidden layer

Input layer						
1st	Node/Neuron	1	2	3	4	5
H	1	3.616	-10.592	6.445	-0.803	7.609
I	2	1.487	-3.305	6.401	-8.857	-7.454
D	3	-6.879	-8.028	1.495	-1.449	-5.830
D	4	8.555	-8.733	-3.917	-1.777	-3.408
E	5	8.187	-0.1361	-5.744	-0.677	-2.193
N	6	2.184	6.143	0.280	-3.938	8.474
L	7	-2.232	2.841	5.439	-0.616	-5.235
A	8	3.212	-5.740	4.520	1.885	2.110
Y	9	0.323	0.703	6.933	-2.306	-1.384
E	10	-4.741	-6.151	-3.604	-2.721	-3.421
R	11	6.175	2.160	-5.064	6.039	4.949

Table 3. Weight matrix (W2): First hidden layer to second hidden layer

		First hidden layer										
2nd	**Neuron**	**1**	**2**	**3**	**4**	**5**	**6**	**7**	**8**	**9**	**10**	**11**
H	**1**	1.72	-5.12	1.13	-1.38	-3.55	4.65	1.29	-1.73	-2.81	-0.80	1.54
I	**2**	5.27	-2.70	-6.95	0.04	-1.55	0.64	3.20	-0.21	-2.88	2.18	2.03
	3	-0.33	-1.10	0.28	-1.64	1.04	3.78	1.09	-0.72	-3.27	-1.80	3.21
D	**4**	1.34	2.85	-1.30	-1.22	2.69	-0.03	-0.70	5.07	-2.47	-1.25	-0.51
D	**5**	5.57	2.95	-0.08	-4.12	0.41	3.43	-1.64	3.82	-5.41	3.67	-0.07
E	**6**	1.34	-2.59	3.80	1.01	1.87	-0.06	0.22	-3.79	1.80	1.00	-5.07
	7	1.59	3.46	-4.04	-2.75	1.36	1.70	-4.23	0.85	0.13	-3.52	1.17
N	**8**	2.30	1.19	-5.73	-3.71	2.69	-2.45	-1.88	-1.41	7.57	-0.03	-1.71
L	**9**	-1.55	-2.07	-2.96	5.93	1.45	-2.18	0.01	-2.96	-3.19	-3.10	1.57
	10	1.80	0.30	-4.57	0.17	1.34	-0.90	-5.42	2.71	2.09	-1.33	2.05
A	**11**	-3.02	-1.36	-0.74	-2.13	2.56	-3.69	2.32	-2.96	0.75	0.73	0.88
Y	**12**	3.64	-0.33	-1.61	-0.98	1.48	1.74	4.68	2.88	0.65	-0.43	1.13
E	**13**	-0.88	-4.18	3.21	-0.15	-0.04	-5.07	0.12	-3.27	2.15	-3.42	0.57
	14	1.59	-2.93	2.45	-1.59	0.63	-2.93	0.23	1.10	0.25	0.17	-2.66
R	**15**	1.63	-3.34	2.74	1.97	1.08	-0.91	-1.14	2.35	0.24	2.65	-2.66

Table 4. Weight matrix(W3): Second hidden layer to output layer

	Second hidden layer								
Node	**1**	**2**	**3**	**4**	**5**	**6**	**7**	**8**	**9**
1	0.083061	-0.31782	0.68952	-0.20642	4.8646	0.25341	0.013381	0.061561	-0.09534
2	-0.11615	-0.16453	-0.45298	0.15566	-0.8753	0.19743	-0.0619	0.048504	-0.08604
3	-0.0079	-0.17943	0.79813	0.12703	0.84186	-0.02721	-0.09654	-0.03391	-0.05348
	10	11	12	13	14	15			
1	0.26627	1.7979	0.075096	0.036696	-0.26828	0.061209			
2	0.22826	-0.62047	0.24927	0.16899	-0.04903	-0.97519			
3	0.045392	0.48791	-0.07892	-0.16402	-0.00634	0.11095			

Table 5. Bias vectors

Neuron	First hidden layer (B1)	Second hidden layer (B2)	Output layer (B3)
1	-8.3068	-1.1669	-0.67829
2	1.1818	-3.051	1.2677
3	9.8091	1.2938	-0.65573
4	0.98752	-2.4466	
5	0.24481	-6.504	
6	-5.7171	-0.4461	
7	-1.0265	-0.12883	
8	-3.0099	-1.2248	
9	3.0187	3.0971	
10	6.3441	1.3272	
11	0.50787	-0.36387	
12		-2.7288	
13		1.5206	
14		5.4298	
15		5.0039	

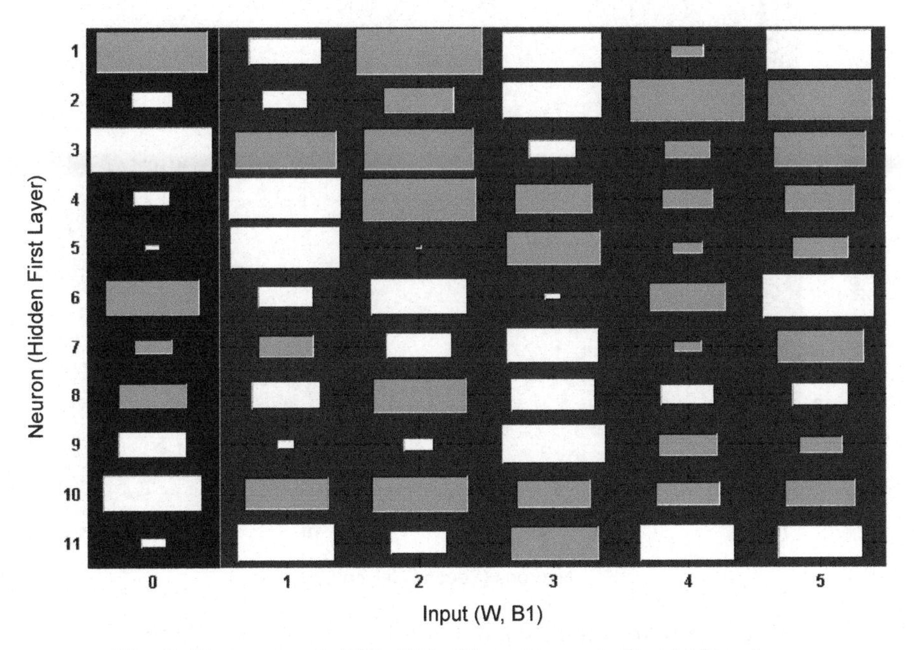

Fig. 6. Hinton graph (W1, B1) of input layer to first hidden layer

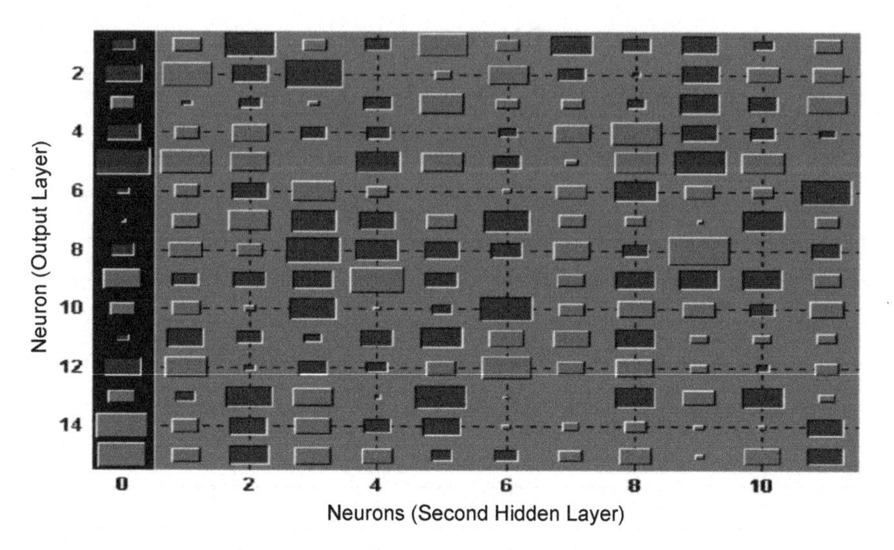

Fig. 7. Hinton graph (W3, B3) of first hidden layer to second hidden layer

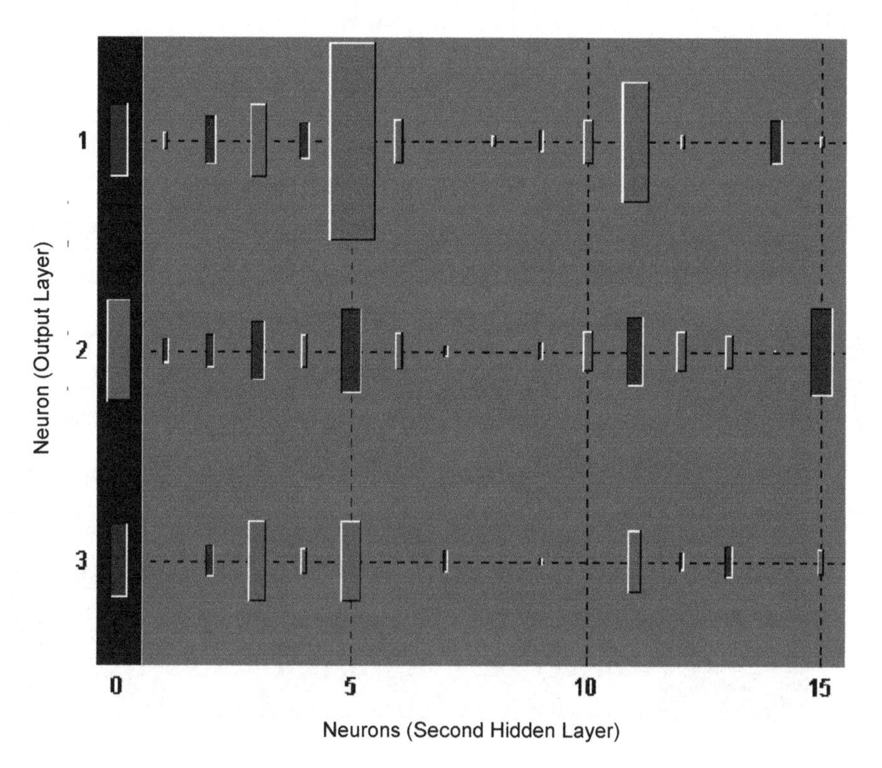

Fig. 8. Hinton graph (W3, B3): Second layer to output layer

Performance Evaluation Measures

The performance indices used for checking the mapping accuracy of the ANN model for various architectures are Absolute Relative Error (ARE), Mean Absolute Relative Error (MARE) and Correlation Coefficient as shown below.

1. Absolute Relative Error (ARE)

$$\text{ARE}(\%) = \frac{100 \times \left|V_p - V_o\right|}{V_p}$$

Where,

$$V_p = \text{Predicted trip rate}$$
$$V_o = \text{Observed trip rate}$$

2. Mean Absolute Relative Error (MARE)

$$\text{MARE}(\%) = \Sigma ARE/N$$

Where,

$$\Sigma ARE = \text{Total ARE of all samples}$$
$$N = \text{Total number of samples}$$

3. Coefficient of Correlation (R)

$$R = \frac{\sum_{i=1}^{N}\left(Q^i_{pre} - \bar{Q}_{pre}\right)\left(Q^i_{obs} - \bar{Q}_{obs}\right)}{\sqrt{\left[\sum_{i=1}^{N}\left(Q^i_{pre} - \bar{Q}_{pre}\right)^2\right]\left[\sum_{i=1}^{N}\left(Q^i_{obs} - \bar{Q}_{obs}\right)^2\right]}}$$

Where,

$$Q^i_{pre} = \text{Predicted value}$$
$$\bar{Q}_{pre} = \text{Mean predicted value of variable}$$
$$Q^i_{obs} = \text{Observed value of variable}$$
$$\bar{Q}_{obs} = \text{Mean observed value}$$

CASE STUDY 2: TRAVEL TIME PREDICTION USING ANN

Time of travel is an important factor in planning of journey either within the city or between the city. It helps the traveller in taking proper decision regarding time of departure from origin as well as maintaining a time buffer so that he can reach the destination in time. Tavel time estimation on different links of the corridor helps traffic planner in assigning the traffic on these links according to the estimated link times for different period of a day. For the present study, an arterial corridor located in the South-West zone of Surat metropolitan area is considered. The corridor has two connected segments: 2.2 km long three lane dual carriageway and 1.1 km long two lane dual carriageway. Traffic volume and speed data was collected during morning peak duration and evening peak duration on a typical weekday.

ANN ARCHITECTURE

Since travel time is a continuous variable, the first modelling formulation used was a Multiple Linear Regression (MLR) with travel time of passenger cars as the dependent variable. In this case independent variables that were included in the model were vehicle compositions of different modes that is per cent 2W, per cent 3W, percent 4W and total volume of vehicles in the traffic stream present on the road section. Multiple linear regression analysis did not yield reasonable parameter estimates and model fit hence the ANN technique was used for the modelling purpose. Methodology explained in the case study 1 has been followed here. Overall network architecture is decided (Fig. 9) based on the performance of the network during different trials and the one which has the minimum number of nodes in the hidden layer Figure 10 shows the average of minimum RMSE of each trial with increment of hidden neurons in the hidden layer. Initially, three hidden neurons were considered and incremented

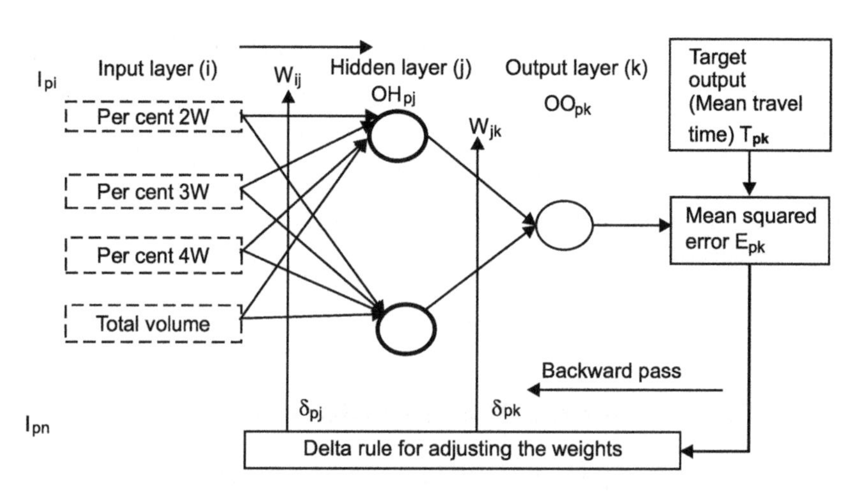

Fig. 9. Structure of the ANN model for modelling

two hidden neurons upto a maximum of 20 hidden neurons. It is seen that the RMSE decreases by increasing the number of hidden neurons in the hidden layer up to a certain extent. Less number of neurons will cause inadequacy in mapping between input and output variables while more number of neurons in the hidden layer may be responsible for over fitting of the model.

Based on the results of the sensitivity analysis, 18 and 20 hidden nodes (Fig. 10) are considered optimum for mapping between input and output variables in case of the present problems. The optimum neural network with 4 input nodes (composition parameters), 18 hidden nodes and 1 output node (average travel time) is further used to study the adequacy of the training and testing for valid generalization of two-lane arterial composition model whereas 20 hidden nodes and 1 output node (average travel time) is used to for valid generalization of three lane arterial road composition model. Data consisting of 120 samples were considered in the model development and validation of both two-lane arterial composition model as well as of three-lane arterial composition model. The data sets are divided into ratio of training (70%) and testing (30%) while running the ANN models.

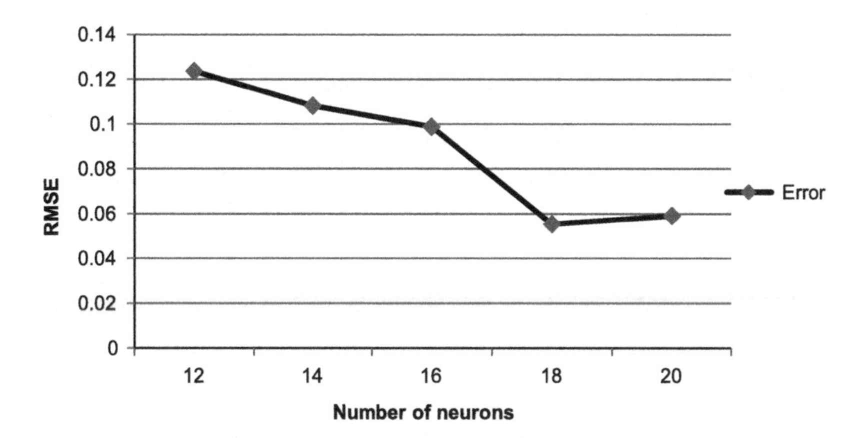

Fig. 10. Estimated optimum hidden nodes for two lane composition model

Performance of ANN model During Training and Testing

Out of 120 observation records 80% samples (96) were considered for model training and remaining 20% (24) samples were considered for model testing and the optimized neural network discussed in the earlier section was considered and sigmoid activation function was taken both input layer activation and hidden layer activation. Thousand iterations were considered to train the network. The performance parameters of two-lane composition ANN model while training and testing are tabulated in Table 6. Statistical performance measures i.e. t-statistic and RMSE for different neurons are presented in Table 7. Figure 10 represents the performance of two-lane composition ANN model at different neurons. Figure 11 shows the accuracy of ANN model for two-lane study corridor. The value of coefficient of determination observed is 0.99 for two-lane composition ANN model.

Table 6. Performance of BPNN two lane composition model during training and testing

Three-lane composition ANN model	Training	Testing
Number of input neurons (Total volume and composition variable)	4	
Number of hidden neurons	18	
Number of output neurons (Average travel time)	1	
Sample size	120	24
R^2 value	0.99	0.760

Table 7. Statistical performance of two-lane composition ANN model

Trial number	Number of neurons	t-statistic	RMSE
1	12	-0.01274	0.1236
2	14	-0.03303	0.1081
3	16	-0.08683	0.0988
4	18	0.011639	0.0553
5	20	0.01700	0.05910

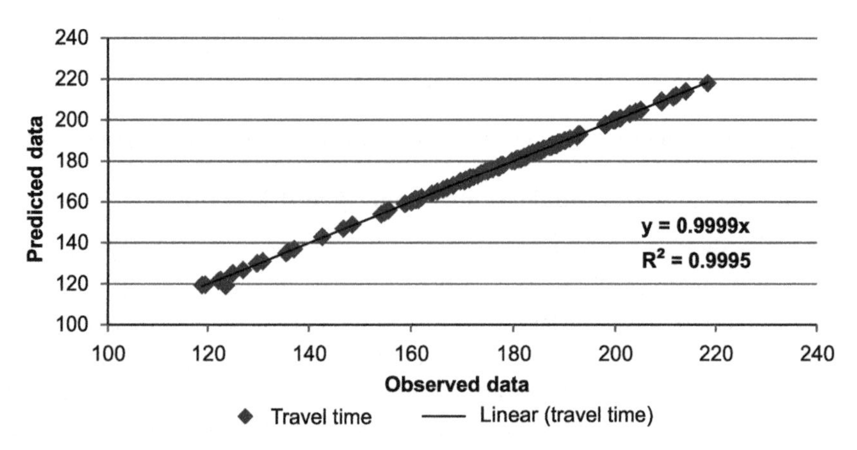

Fig. 11. Observed and estimated travel time for two lane data during testing

The performance parameters of three lane composition ANN model while training and testing are tabulated in Table 8 however statistical performance measures i.e. t-statistic and RMSE for different neurons are presented in Table 9. The Figure 12 represents the performance of three-lane composition ANN model at different neurons. It can be observed that the value of RMSE is decreasing as per the increase in neurons. In case of 20 neurons the R^2 value was highest and lowest amount of RMSE was noted hence the results are sealed and weights are noted down for the further analysis part.

Table 8. Performance of three-lane composition ANN model

Three-lane composition ANN model	Training	Testing
Number of input neurons (Total volume and composition variable)	4	
Number of hidden neurons	20	
Number of Output Neurons (Average travel time)	1	
Sample size	120	24
R^2 value	0.98	0.620

Table 9. Statistical Performance of three-lane composition ANN model

Trial number	Number of Neurons	t-statistic	RMSE
1	12	-0.06874	0.1729
2	14	-0.06168	0.1462
3	16	-0.02269	0.1352
4	18	0.003756	0.0867
5	20	0.006177	0.0464

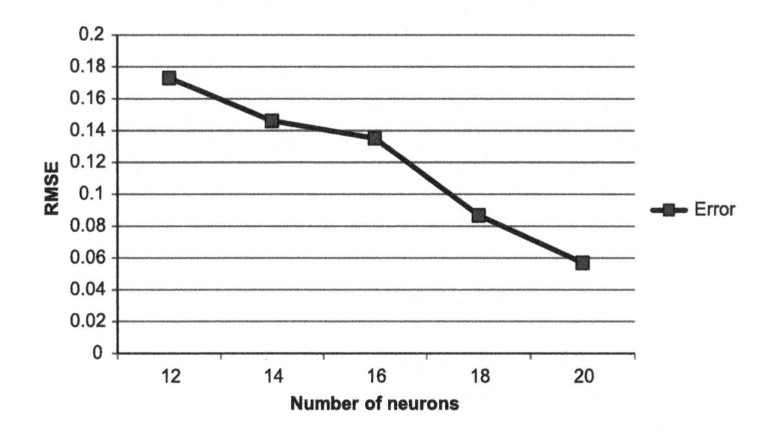

Fig. 12. Estimated optimum hidden nodes for three-lane composition model

Figure 13 shows the accuracy of ANN model for three-lane study corridor. The value of coefficient of determination observed is 0.99 for three-lane composition ANN model.

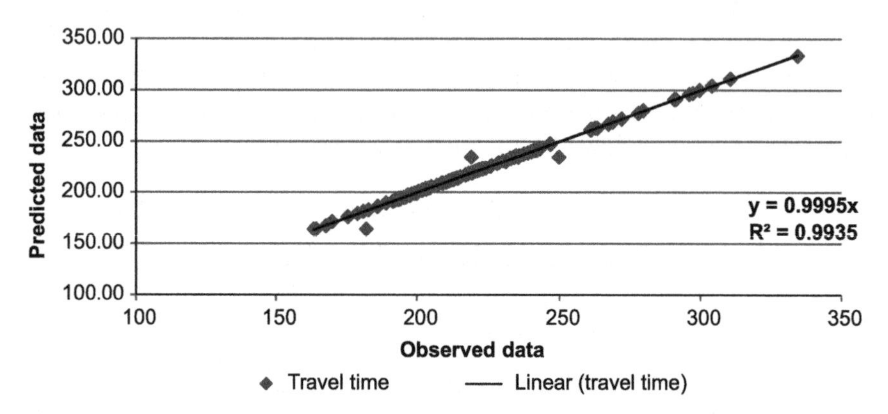

Fig. 13. Observed and estimated travel time for three-lane data during testing

CONCLUSION

The technique of artificial neural network is a model free modelling technique that does not require any presumptions about either the nature of the variables and their distributions or nature of relationship between explained and explanatory variables. This is demonstrated through two case studies related to transport planning and traffic engineering. The first case study of travel demand estimation shows how visibly correlated explanatory variables can be used effectively to predict travel demand for different zones of a metropolitan city and outputs can be further applied to predict change in urban form. The second case study illustrates how statistically validated ANN models can be developed for travel time prediction for an urban corridor considering traffic flow and composition. Both the models have exhibited very good prediction accuracy and hence can be very useful for application in transportation engineering practice.

REFERENCES

Anderson, Michael, D. and Olander, Justin P. (2002). Evaluation of Trip Generation Techniques for Small Area Travel Models, *Journal of Urban Planning and Development,* 128(2),pp. 77-88, June.

Jang, J.S.R., Sun, C.T. and Mizutani, E. (1997). *Neuro-Fuzzy and Soft Computing*. Upper Saddle River, N.J.: Prentice Hall.

Jeng, D.S, Cha, D.H. and Blumenstein, M.(2003). Application of Neural Network in Civil Engineering Problems, *Proceedings of the International Conference on Advances in the Internet, Processing, Systems and Interdisciplinary Research,* Griffith University Gold Cost Campus, Australia, July 20.

Joshi, G.J. and Katti, B.K. (2004). Modelling Zonal House hold Growth Rate in Metropolitan Context, *National Conference on Urban Housing Issue and Development strategies*, SVNIT, Surat, 1,pp. 19-29.

Kalenoja, Hanna. (1999). Spatial Differences in the Trip Generation and Travel Behaviour: Empirical Observation in the Tampere Region, *Urban Transport Systems Conference in Lund,* June 7.

Kumar, Ajay and David, Levinson. (1992). "Specifying, Estimating and Validating a New Trip Generation Model: Case Study in Montgomery County, Maryland". *Transportation Research Record,* 1413, pp. 107-113, December 9.

Liu, H., Vent Lint, H. and Zuylen, H. (2006). *Neural Network Based Traffic Flow for Urban Arterial Travel Time Prediction*. Faculty of Civil Engineering and Geosciences, Delft University of Technology.

Xie, Chi, Lu, Jinyang and Parkany, Emily. (2003). Work Travel Mode Choice Modeling Using Data Mining: Decision Trees and Neural Network, *Transportation Research Record*, pp. 12-16, Washington, January.

Genetic Algorithm and its Applications on Water Distribution Networks

Rajesh Gupta[1]

ABSTRACT

In the present paper, the application of genetic algorithm (GA) had been applied water distribution network. The population size and number of generation are taken as 100 and 70, respectively.

Keywords: *Genetic algorithm, water distribution network*

INTRODUCTION

Genetic algorithm is a search-based optimization tool in which a complex solution space is systematically searched through successively generating improved solutions (Goldberg, 1989). The approach has been derived from famous Darwin philosophy of "survival of fittest". It distinctly differs from traditional methods as it works with a coding of parameter set and not on the parameters. It searches from population of points and not from a single point, and uses information regarding objective function and not a derivative or other auxiliary knowledge and it uses the probabilistic transition rules.

Genetic algorithm searches the solution space by creating some initial population (i.e. Number of solutions) in the search space, selecting them in the next evolution according to their fitness and combining them using certain mechanisms. This process of selection, evaluation and combination continues till no further improvement in fitness is observed; or a specific number of generations are reached. It is observed that the number of solutions to be tested for hydraulic simulations is much less as compared to that needed for complete enumeration. The efficiency and effectiveness of GA depends on several parameters and suggestions have been made by several researchers to improve GA. A flow diagram of basic GA process is shown in Figure 1.

1. Professor of Civil Engineering, Visvesvaraya National Institute of Technology, Nagpur, 440 011, India, E-mail: drrajeshgupta123@hotmail.com

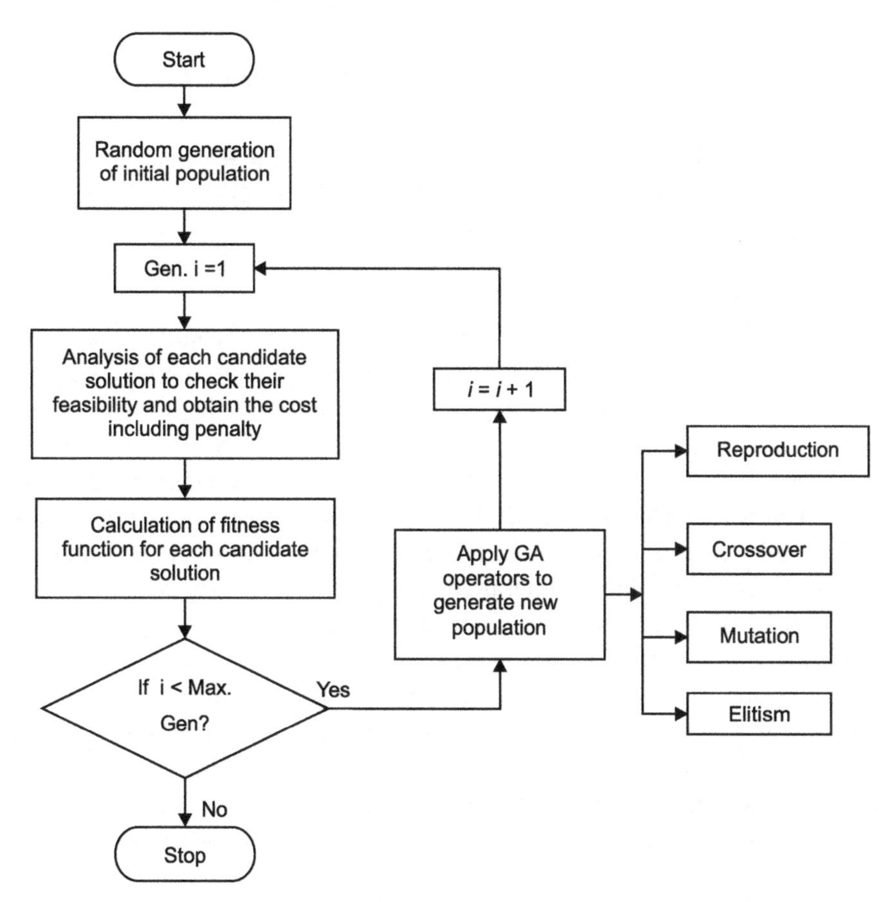

Fig. 1. A flow diagram of simple GA

GA APPLICATION TO DESIGN OF WATER DISTRIBUTION NETWORKS

The general optimization problem for a gravity WDN can be expressed as (Dandy et al. 1996): For a given layout of pipes governed by topography and road network and a set of specified demand patterns at the demand nodes, find the combination of pipe sizes which gives the minimum material and construction cost subjected to constraints of node flow continuity, loop head loss, maximum and minimum head, minimum and maximum diameter of links.

GA handles constraints by adopting penalty to objective function. Therefore, the objective function for optimal design of a WDN is modified using the penalized function as

$$\text{Minimize} f(D_1, \ldots, D_X) = \sum_{x=1}^{X} c(D_x) \times L_x + \sum_{j=1}^{M-S} p_j \times \{\max(H_j^{\min} - H_j, 0)\} \quad (1)$$

Where, $c(D_x)$ = unit cost of pipe having diameter D; p_j = penalty multiplier at node j; $\{\max\ (H_j^{\min} - H_j,\ 0)\}$ = violation of the pressure constraint at node j. Penalty is imposed for violation of each pressure constraint. Several modifications have been suggested to GA to improve its efficiency in obtaining optimal solution.

High crossover probabilities and low mutation probabilities have been suggested for better performance of GA (Goldberg and Kuo 1987, Simpson et al. 1994 Daniel et al. 2007). Gray coding is found to be better than binary coding for string representation (Dandy et al. 1996). A real coded string is recommended over binary and gray coding to avoid the problem of redundant states (Vairavamoorthy and Ali 2000). A real coded string does not require coding and decoding and therefore reduces computational time. Daniel et al. (2007) suggested alpha-numeral coding instead of binary coding. Halhal et al. (1997) suggested a structured messy genetic algorithm using a variable-length string representation for the rehabilitation of WDNs. Wu and Simpson (2001) proposed a messy genetic algorithm for the optimization of WDNs. Dandy et al. (1996) used a variable power scaling of the fitness function and adjacency mutation operator to improve GA. Savic and Walters (1997) used a graded fitness function, and rank selection. Savic and Walters have attempted the problem as a least cost optimization problem with only pipe diameter being decision variables. Other variables especially pipe layout are considered known. Vairavamoorthy and Ali (2000) proposed a regression model to prime the network solver, to reduce the number of iterations required by the solver. Wu and Walski (2005) recommended a self-adaptive penalty approach over other constraint-handling methods for WDN optimization. Computational extensiveness is a major drawback with GA. Lippai et al. (1999) observed a linear relationship between the population and the iterations required for the least-cost solution and suggested a GA search initially with a small population. Then, repeating the GA search with double the population. If the least-cost solution does not improve after doubling the population twice, there is a high probability that the optimal solution is found. Balla and Lingireddy (2000) suggested parallel computing on a network of PCs to reduce the computational time for the calibration of WDNs. Kumar et al. (2006) used cluster of Workstations for redundancy-based design of WDNs.

For practical large networks, GA search across huge discrete space may lead to a delayed convergence. Thus, it reduces the possibility of getting the global optimum while increasing the computational time. By reducing the search space it is possible to generate feasible solutions more quickly and obtain an optimal solution in a fewer number of generations resulting in a substantial saving in terms of computational time. Vairavamoorthy and Ali (2005) used a pipe index vector to improve the GA-based pipe optimization. Kadu et al. (2005, 2008) suggested use of critical path method for initially identifying a set of candidate pipe sizes for each link, thereby reducing the search space drastically, and GA search in reduced search space for reducing the computational time.

The modified GA optimization of Kadu et al. (2008), is based on the following:

Representation Scheme: Integer-coding scheme, which has the advantage that discrete diameters are directly used to form solution-strings, is adopted. The coding and decoding of the actual variables is not required, thus saving computational time.

Penalty Multiplier: Capitalized energy cost is used as a penalty multiplier. The penalty multiplier at node j, p_j, is accordingly obtained considering the rate of interest of 8%, design life of 30 years, power factor = 0.6 and energy rate of Rs. 4.5 per kWh.

Fitness Function: Scaled fitness function given by Dandy et al. (1996) is used. This function is given by some power (n) of the raw fitness of a solution f_i. The variable exponent (n) is used to modify fitness of the function. The value of n is allowed to increase in steps as the GA run proceeds. A low value of n (= 1) is employed during the initial runs; and is increased to 3 or 4 during later generations.

Reproduction Operator: Tournament selection scheme is used as reproduction operator. In tournament selection a pair of solutions is selected from the population and the best one is retained. By repeating the process for specified number of times, a mating pool of suitable size is formed which will contain multiple copies of good potential solutions and a few or no copy of less fit solutions.

Cross Over: Cross over operator, which creates new off-springs after selection by partially exchanging information among parent-solution strings is an important operator in GA search. Basic crossover operators (one-point, two-point, uniform) have been used by different researchers, in isolation or otherwise, in previous studies. Eshelman and Schaffer (1993) suggested blend cross over operator in which two-parent strings combine to form a child string. Ono and Kobayashi (1997) and Kita et al. (1998) suggested multi-parent cross over operator with three or more parents. Multi-parent and universal parent crossovers are used along with basic crossover operators randomly (Goldberg 1989, Michalewicz 1992, Deb 2001, Kadu et al. 2008). In multi-parent crossover three or four parent strings are selected from the mating pool formed after reproduction. One is a mother string, which produces an offspring after crossover with the remaining parent strings. In universal-parent crossover, mother string selected from the mating pool combines with universal parent (i.e., the entire set of commercially available pipes) to produce a new off-spring but with a very less probability (<= 5%).

In all, nine types of cross over operators—simple uniform, simple single point, simple two point, multi-parent uniform, multi-parent single point, multi-parent two-point, universal parent uniform, universal single-point and universal two-point—are used. Initially, a mother string is selected with cross over probability of 0.95. Next, random number from 1 to 9 is generated to select type of cross over operator. Depending upon the selected type of operator, other parent strings are selected randomly and offspring is produced.

Mutation: In the initial few generations selection and cross over play an important role while in the later generations crossover and mutation play a significant role. Therefore, instead of using uniform mutation wherein mutation probability, p_m, is kept constant in the search process, non-uniform mutation is used in this study wherein p_m increases from a very low value to a maximum as the search proceeds. The mutation rate is usually set very low, e.g., $p_m \in (0.01, 0.05)$ (Simpson et al. 1994). On an average, only one gene (i.e., one pipe size) is changed in each offspring. In the present work minimum and maximum probability of mutation is considered as 1% and 10% respectively.

Simple and universal mutation use a set of candidate diameters for a link and an entire set of commercially available pipes respectively. The neighbour mutation uses one or two pipe sizes on higher or lower side of a size of link selected for mutation. This arrangement facilitates: (1) Consideration of the pipe sizes, depending on their relative importance in the optimization process; and (2) Systematic and diversified search through the entire search space.

Elitism: In the process of GA search, better and better solutions start emerging in the population. However, the best and a few better solutions in the population may get lost if they get subjected to GA evolution process. Elitism operator preserves the best and a few better solutions from a current population for the next population.

Hydraulic Simulation: GA handles the constraints externally using the hydraulic solver. The use of an external hydraulic solver based on iterative methodology coupled with the large number of GA evaluations makes the process computationally exhaustive. EPANET (Rossman 2000) based on the same methodology can be used.

Reduction in Search Space: It is the most important parameter that controls the efficiency and effectiveness of a GA search, especially for large practical networks to be designed with many commercial available pipes. The solution space can be substantially reduced using critical path method of WDN as explained with the example later on.

GA methodology: The GA methodology consists of the following steps:

1. Using path concept converts the looped network to a tree; and classifies links as primary or secondary.
2. Using critical path concept, obtain the HGL values at the intermediate nodes.
3. Using these HGL values, determine flow in secondary links considering them to be of minimum size. Then, obtain flow in primary links.
4. Obtain the diameter of primary links.
5. Obtain the candidate pipe sizes of each link.
6. Generate initial population of suitable size consisting of pipe sizes selected randomly from the set of candidate pipe sizes.
7. Compute network cost for each solution in the population.
8. Carry out hydraulic analysis of each network to check feasibility of solutions.
9. Compute penalty cost for infeasible solutions.
10. Compute total cost of each solution.
11. Compute fitness of each solution.
12. Carry out reproduction to obtain new population.
13. Apply crossover, mutation and elitism operators to generate new solutions.
14. Repeat steps 7 to 13 for specified number of generations.

ILLUSTRATIVE EXAMPLE

The entire GA methodology, including selection of candidate diameters of each link using critical path concept, is explained with illustrative example. A single source, 7-link network consisting of five demand nodes, is shown in Fig. 1(a). Node 1 is a source node with constant HGL of 100 m. The demand nodes are labelled 2 through 6, the minimum required HGLs in metres and nodal demands in cubic metre per minute are shown in the figure. The links 1 through 7 (underlined) and their lengths in meters (given in parenthesis) are shown along the links. The commercially available pipe sizes in mm and their unit cost in rupees shown in parentheses are: 150 mm (Rs. 1115); 200 (1600); 250 (2154); 300 (2780); 350 (3475); 400 (4255); 450 (5172); 500 (6092); 600 (8189); 700 (10670); 750 (11874); 800 (13261); 900 (16151); and 1,000 (19395). The Hazen-Williams coefficient is 130 for all links. The network is designed using modified GA with reduction in search space. GA operators are taken as: population size = 60; p_c = 0.95; p_m = 0.02 to 0.05; and number of generations = 25.

Paths to different nodes and the path lengths to all demand nodes are obtained in Table 1; and are also shown in Fig. 1(b). The primary and secondary links are identified and are shown by thick and thin lines, respectively. Links 1 to 5 are primary links, and links 6 and 7 are the secondary links. In Table 1, the critical path and the critical sub-paths, the critical slope and the critical sub-slopes (underlined in Table 1) are determined. The H_j; $j = 2, \ldots, 6$ are estimated and are shown in Fig. 1(b). In Eq. 1, α and β are taken as 1.85 and 4.87, respectively. For h_f and L in meters, Q in cubic meters per minute, and D in millimeters, ω is 2.234×10^{12}. Using Eq. 1 and taking minimum specified diameter = 150 mm for secondary links 5 and 6, discharges in them are estimated. Using node flow continuity equation, the discharges in the primary links are also estimated (Col. 2, Table 2). Considering pipe discharges and HGL values obtained using critical path, required continuous size of primary links are obtained (Col. 4, Table 2) and is used to form a set of candidate diameters for all links (Col. 5, Table 2).

With 14 commercially available pipe sizes the possible solutions are 14^7 (= 105,413,504). Modified GA reduces the search space to 5^7 (= 78,125, i.e. 0.074% of the entire search space). Because of the repetition of the smallest available pipe size, the number of possible solutions is further reduced to $3^2 \times 4^2 \times 5^3$ (= 18,000, i.e. 0.017% of entire space).

Modified GA provided the solution costing Rs. 44,48,250 after 1500 evaluations in 9 seconds on P3/ 166 MHz / 32 MB RAM. The link numbers and pipe diameters in millimeters given in parenthesis are: 1 (400 mm); 2 (300); 3 (200); 4 (300); 5 (200); 6 (150); and 7 (150). Node numbers and available HGL at different nodes given in parenthesis are: 2 (96.615 m); 3 (95.872); 4 (94.167); 5 (93.722); and 6 (90.785).

GA APPLICATION FOR MULTI SOURCE WATER TRANSMISSION SYSTEM

In water stressed regions the water allocation form the sources (like dams, rivers) are restricted and rationalized for different uses. Under such circumstances the water transmission system are developed from various sources for respective fixed allocations and of course for the respective governing levels. The problem covers selection of optimal route, optimal allocation of available water from these sources to the demands of service reservoirs and optimal pipe sizing for the selected routes together.

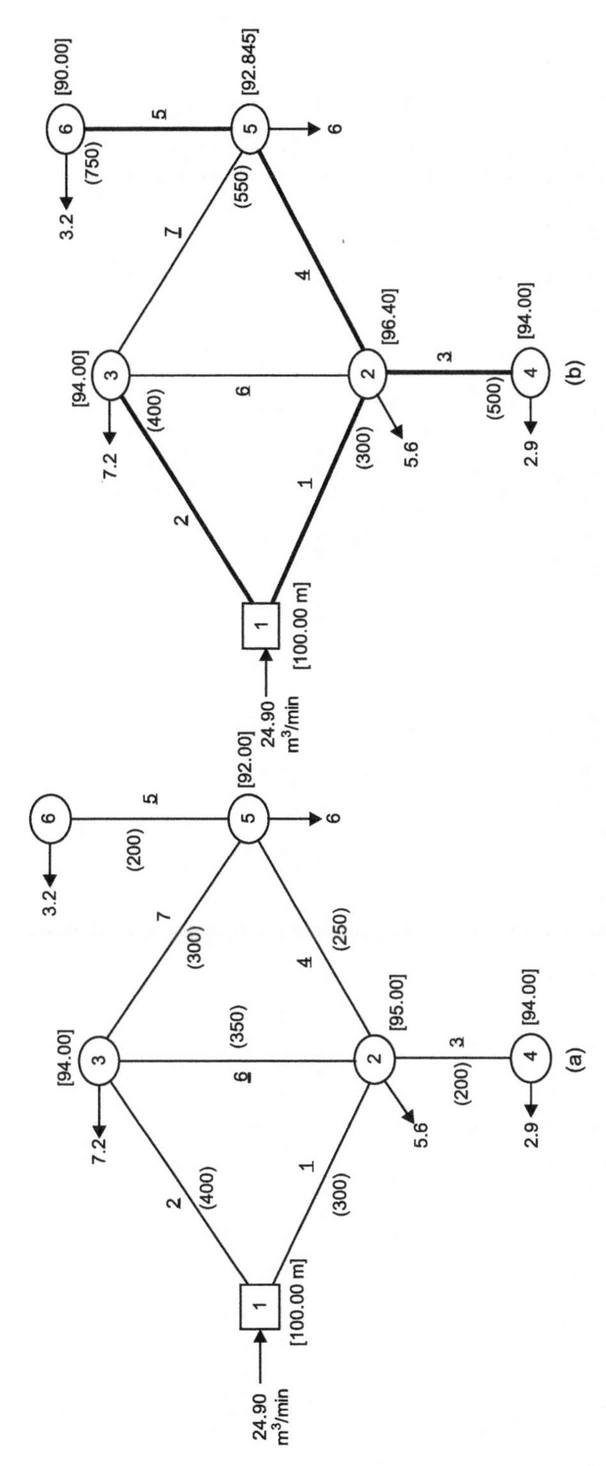

Fig. 1(a). A Single Source Water Distribution Network; (b) Primary and Secondary links using path concept and fixing HGL values using critical path concept

Table 1. Determination of critical paths and critical sub-paths

Distribution tree or sub-tree	Path or sub-path serial number	Path	Length of path (m)	HGL (m) at the Source of path	HGL (m) at the End of path	Maximum available friction loss (m)	Slope of path
(1)	(2)	(3)	(4)	(5)	(6)	(7)	(8)
Distribution tree							
1	1	1 - 2	300	100.00	95.00	5.00	0.01667
	2	1 - 3	400	100.00	94.00	6.00	0.01500
	3	1 - 2 - 4	500	100.00	94.00	6.00	0.01200[a]
	4	1 - 2 - 5	550	100.00	92.00	8.00	0.01454
	5	1- 2 - 5 - 6	750	100.00	90.00	10.00	0.01333
Distribution sub-tree							
1.1	1	1 - 3	400	100.00	94.00	6.00	0.01500[a]
	2	2 - 5	250	96.40	92.00	4.40	0.01760
	3	2 - 5 - 6	450	96.40	90.00	6.40	0.01422[a]

[a]critical slope (minimum slope) is shown underlined.

Table 2. Candidate diameters using path concept and critical path method

Link number	Link discharge (m3/min)	Head loss in link (m)	Diameter (mm)	Candidate diameters (mm)
(1)	(2)	(3)	(4)	(5)
1	12.366	3.60	348.37	250, 300, 350, 400, 450
2	6.934	6.00	267.11	150, 200, 250, 300, 350
3	2.900	2.40	200.81	150, 150, 200, 250, 300
4	8.472	3.56	291.41	200, 250, 300, 350, 400
5	3.200	2.85	201.30	150, 150, 200, 250, 300
6[a]	0.994	2.40	-	150, 150, 150, 200, 250
7[a]	0.728	1.16	-	150, 150, 150, 200, 250

[a]Secondary links.

Consider that a WTS is to be designed to cater water to various service reservoirs in a city from various identified sources. These sources are having fixed water availability and are at fixed hydraulic gradient levels. The layouts have been specified from which feasible routes can be identified to transmit the water from these sources. The demands and HGL of the service reservoirs are known. Now the problem in its simple term has following components:

a. The demands of the service reservoirs should be allocated in a optimal way with respect to water availability and HGL to the identified sources.
b. The proper routes should be selected from the geometrical layout to transmit the water from these sources to the service reservoirs with respect to the allocation in step a.
c. The most economical diameters to be selected for the stretches selected in step b.

ILLUSTRATIVE EXAMPLE

Consider a WTS layout shown in Figure 2 to illustrate the application of the proposed methodology. The network consists of 3 source node, 7 service reservoirs and are having 25 possible pipe stretches. The link details are given in Table 3. The cost of available pipe diameters is presented in Table 4. The node details are given in Table 5 and source details are given in Table 6.

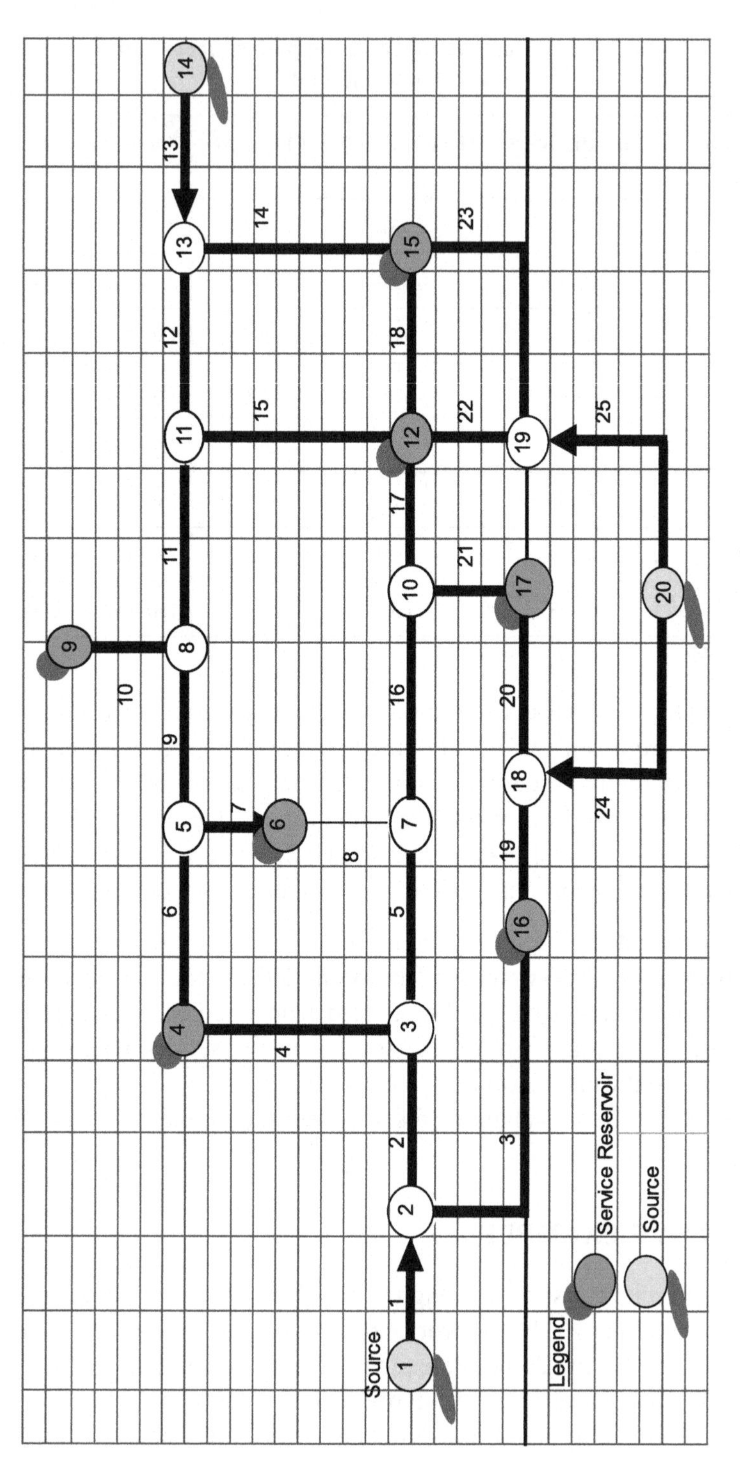

Fig. 2. A water transmission network

Table 3. Network information (link details)

Link No.	Length	Link No.	Length	Link No.	Length	Link No.	Length	Link No.	Length
1	3500	6	1600	11	2000	16	2000	21	800
2	1800	7	800	12	1500	17	2000	22	600
3	1000	8	200	13	500	18	1500	23	3000
4	1200	9	2000	14	1000	19	400	24	2600
5	2500	10	400	15	400	20	1500	25	1000

Table 4. Network information (cost data)

Diameter, mm	80	100	150	200	250	300	350	400
Cost Rs/m	570	627	882	1211	1617	1966	2350	2824
Diameter, mm	450	500	600	700	750	800	900	1000
Cost Rs/m	3386	3971	5297	6691	8471	8700	10710	12843

Table 5. Network information (node details)

Node No	Type	Ground level (m)	Full supply level (m)	Demand (MLD)
4	Service reservoir	5	22	20
6	Service reservoir	5	25	40
9	Service reservoir	5	16	10
12	Service reservoir	5	16	45
15	Service reservoir	5	18	60
16	Service reservoir	5	18	35
17	Service reservoir	5	18	30

Table 6 . Network information (source details)

Source at node	FSL (m)	Water available (Mld)
1	80	120
14	70	70
20	60	50

The GA application steps are elaborated as follows:

a. **Initial Population for Individual Phase:** It is decided to generate population of 100 candidate solutions to begin the searching. In each solution for all twenty-five pipes a random number is assigned between 0 to 1. Then the diameter is assigned with respect to this random number. Sixteen commercial size diameters have been considered. The range of the random number and the diameter has shown in Table 7.

Table 7. Diameter assignment with respect to random number

Random No.	Diameter	Random No.	Diameter	Random No.	Diameter
0.0000	80	0.3750	350	0.6875	700
0.0625	100	0.4375	400	0.7500	750
0.1250	150	0.5000	450	0.8125	800
0.1875	200	0.5625	500	0.8750	900
0.2500	250	0.6250	600	0.9375	1000
0.3125	300				

If the random number is in between 0.0000 to 0.0625, the diameter selected is 80 mm.

b. **Evaluation of Solutions:** Once the diameters are assigned to these candidate solutions, then the networks are solved with the help of network solver available with EPANET 2.0. The output of each solution is then noted for water withdrawn from each of the sources and for fulfilling of pressure and flow requirements of each of the service reservoirs.

c. **Penalty Application:** Now for each candidate solution, following penalties are applied

   1. ***Infeasibility penalty***

      This penalty is applied for those infeasible solutions where the pressure requirements are not met. The penalties are worked out with basis of additional energy cost that is required to lift the required demand to the desired HGL.

   2. ***Bad allocation penalty***

      This penalty is applied for not meeting the exact specified withdrawl from the specified sources. Since the whole process is based on random selection, at least in initial round it is not possible that even few solutions shall be meeting the criteria of withdrawing the specified quantity of water from the specified sources. It is observed that in later successive iteration a very minute difference always remains. Therefore this penalty is worked out for all solutions.

   3. ***Excess head penalty***

      This penalty is introduced to encourage the economic selection of diameters. For each solution the excess head available at demand nodes are noted and on the basis of excess waste energy these penalties are worked out.

d. **Grading the Solutions:** For grading the solutions basic cost of the solutions and penalty costs are considered with certain weightages. The solutions are ranked on ascending order with respect to the above total costs. These solutions have assigned with Rank No 100 to 1 with respect to their position. With certain weightages to the total cost and to the rank evaluation cost (fitness) have been worked out with following expression

$$f_i = (1/ C_{evaluation})^s \qquad (2)$$

s = scaling component taken 1 for early generation, suitably increased to higher values.

On the basis of fitness of individual solution, probability for selection in next generation is obtained using the expression

$$P_i = (f_i/\Sigma(f_i)) \qquad (3)$$

e. **Creation of New Population:** The population for next new generation has been created with the following operators:
   1. ***Reproduction:*** Roulette Wheel selection is used for reproduction. The solution having higher fitness is given higher slice on the Roulette wheel based on its probability and population is reproduced.
   2. ***Cross over:*** In the presented application one-point crossover has been adopted. It is performed by randomly choosing a crossing stretch for two randomly selected solutions and by exchanging all the upsizing information of right side of crossing over stretch to each solution of the pair. The operation is carried out on an MS-Excel platform with cross over probability between 0.6 to 0.8 in different runs.
   3. ***Mutation:*** The probability of mutation is taken between 0.2-0.3 in different runs.
   4. ***Elitism:*** Two best solutions are retained.
   5. ***No Change:*** This operator suggests no operation on the randomly selected solution and carry forward them in next generation. This operator has been used to adhere to natural process of evolution. Herein, 5% solutions are carried forward.

This way new population in next generation is produced. It is then subjected to steps b to e. This process is repeated till satisfactory results are achieved.

f. **Results and Analysis:** With a initial population of 100 solutions, 70 generations are carried out. The final selection of water transmission network from the given geometrical layout is presented in Fig. 4 and the diameter for the selected network is presented in Table 9.

Table 8. Progress over successive generation

	Pipe	Penalty	Allocation from sources		
Gen No.	Cost	Cost	Source 1	Source 2	Source 3
1	2177.2	1341.1	125.4	83.1	31.5
10	1815.1	739.0	118.4	72.7	49.0
20	1735.8	452.0	118.9	69.2	51.9
30	1510.7	534.5	119.7	65.5	54.8
40	1415.4	187.8	120.1	70.8	49.2
50	1405.9	162.9	119.8	70.7	49.5
60	1249.1	85.1	120.2	70.3	49.5
61	1245.7	87.2	120.7	69.4	49.9
65	1234.0	80.7	120.4	69.5	50.2
66	1233.5	80.7	120.4	69.5	50.2
70	1239.1	69.4	120.2	69.6	50.3

Table 9. Selected diameter (link wise)

Link No.	Dia	Link No.	Dia	Link No.	Dia	Link No.	Dia	Link No.	Dia
1	900	6	75	11	400	16	450	21	400
2	800	7	75	12	500	17	75	22	300
3	350	8	350	13	450	18	350	23	350
4	300	9	150	14	350	19	75	24	200
5	600	10	250	15	350	20	250	25	500

OTHER APPLICATIONS

GA is best suited for discrete optimization problem. Several applications for facility location problems, in which locations of valves, tanks, booster chlorination, online sensors etc. have been decided by GA. Also, the multi-objective problems of network design considering reliability, redundancy, uncertainty in parameter through probabilistic and fuzzy representation have been solved using GA.

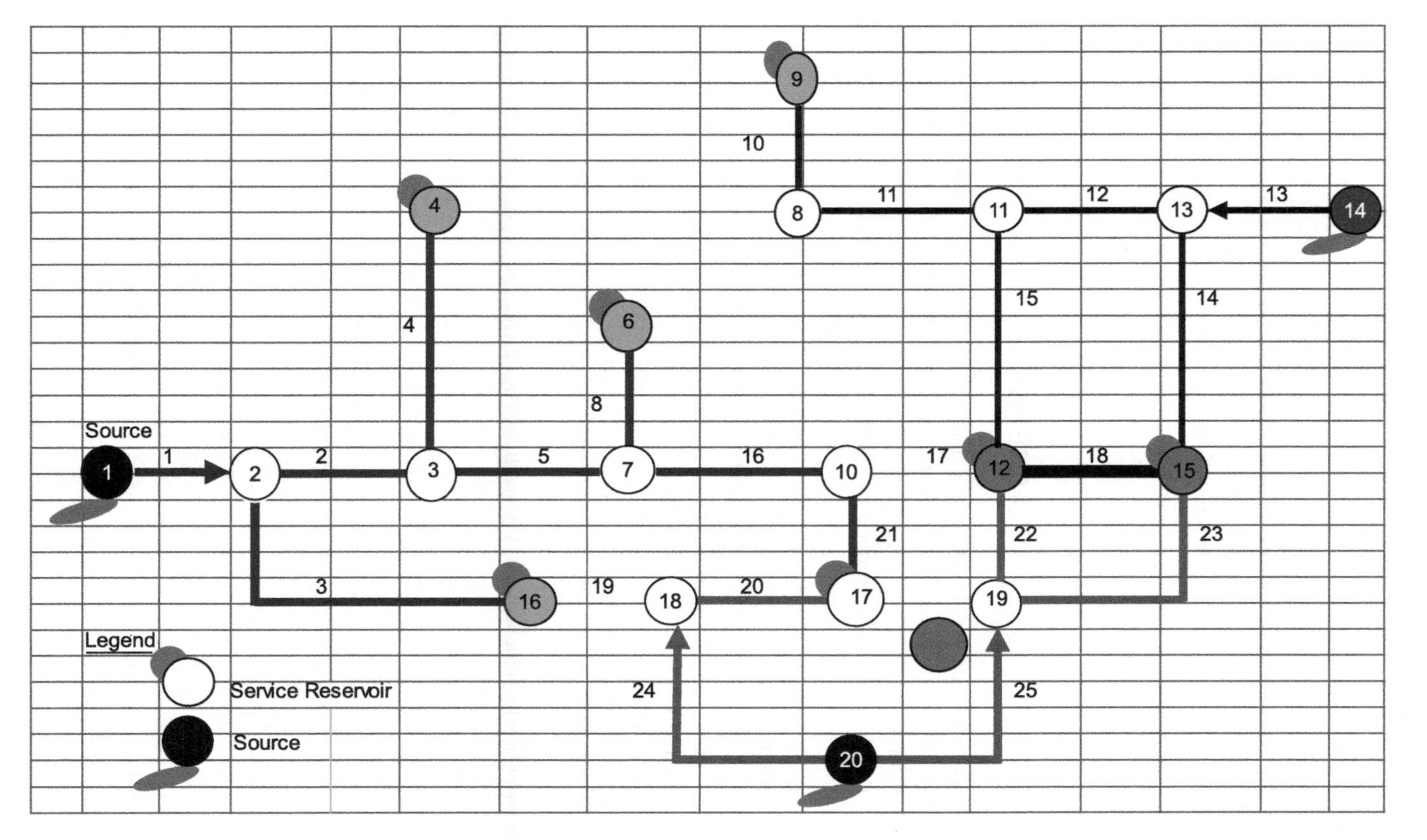

Fig. 3. Selected network from the given geometrical layout

SUMMARY AND CONCLUSION

Genetic algorithm in general has been described and its application to design of water distribution network and water transportation from different sources to different service reservoirs has been seen. The applications of GA to water distribution networks are many. The GA requires large number of simulations and thus computationally extensive. Several methods have been suggested to improve the efficiency of GA. The present trend is to couple GA with other techniques to solve large network problems.

REFERENCES

Balla, M.C. and Lingireddy, S. (2000). Distributed Genetic Algorithm Model on Network of Personal Computers. *Journal of Computing in Civil Engineering, ASCE,* 14 (3), pp. 199-205.

Bhave, P.R. (1978). "Noncomputer Optimization of Single-Source Networks." *Journal of Environmental Engineering, ASCE,* 104(EE4), pp. 800-814.

___________. (2003). *Optimal Design of Water Distribution Networks.* Alpha Science International Ltd.: Pangbourne, England.

Carlson, S.E. (1995). "A General Method for Handling Constraints in Genetic Algorithms." *Proc., Joint Conf. on Information Sci.,* 663-667; also available on http:// vlead.mech. Virginia.edu/publications/paper2.ps.

Dandy, G.C., Simpson, A.R. and Murphy, L.J. (1996). "An Improved Genetic Algorithm For Pipe Network Optimization." *Water Resources Research,* 32(2), pp. 449-458.

Deb, K. (2001). *Multi-objective Optimization using Evolutionary Algorithms.* Wiley: London.

Eshelman, L.J. and Schaffer, J.D. (1993). "Real-coded Genetic Algorithms and Interval Schemata." In: *Foundations of Genetic Algorithms 2 (FOGA-2),* pp. 187-202.

Fujiwara, O. and Khang, D.B. (1990). "A Two-Phase Decomposition Method for Optimal Design of Looped Water Distribution Networks." *Water Resources Research,* 26(4), pp. 539-549.

Goldberg, D.E. (1989). *Genetic Algorithms in Search, Optimization and Machine Learning,* Addison-Wesley: Reading, Mass.

Goldberg, D. E. and Deb, K. (1991). "A comparison of selection schemes used in genetic algorithm". In: *Foundations of Genetic Algorithms 1(FOGA),* pp. 69-93.

Goldberg, D.E. and Kuo, C.H. (1987). "Genetic Algorithm in Pipeline Optimization." *Journal of Computing in Civil Engineering, ASCE,* 1(2), pp. 128-141.

Halhal, D., Walters, G.A., Ouazar, D. and Savic, D.A. (1997). "Water Network Rehabilitation with Structured Messy Genetic Algorithm." *Journal of Water Resources Planning and Management, ASCE,* 123(3), pp. 137-146.

Kadu, M.S., Gupta, R., and Bhave, P.R. (2005). "Optimal Design of Water Distribution Networks using Genetic Algorithm with Reduction in Search Space." *Proc. of*

One-Day National Conference on Geotechniques and Environment for Sustainable Development, Nagpur, India, pp. 182-189.

Kita, H., Ono, I., and Kobayashi, S. (1998). "The Multi-Parent Unimodal Normal Distribution Crossover for Real-Coded Genetic Algorithms." Tokyo Institute of Technology, Japan.

Maier, H.A., Simpson, A.R., Zecchin, A.C., Foong, W.K., Phang, K.Y., Seah, H.Y. and Tan, C.L. (2003). "Ant Colony Optimization for Design of Water Distribution Systems." *Journal of Water Resources Planning and Management, ASCE*, 129(3), pp. 200-209.

Michalewicz, Z. (1992). *Genetic Algorithms + Data Structures = Evolutionary Programs.* Springer-Verlag New York, Inc.: New York, N.Y.

Ono, I., and Kobayashi, S. (1997). "A Real-Coded Genetic Algorithm for Function Optimization using Unimodal Normal Distribution Crossover." *Proc. of the Seventh International Conference on Genetic Algorithms*, pp. 246-253.

Rossman, L.A. (2000). *EPANET, User's Manual.* U.S Envir. Protection Agency: Cincinnati, Ohio.

Savic, D.A. and Walters, G.A. (1997). "Genetic Algorithms for Least-Cost Design of Water Distribution Networks." *Journal of Water Resources Planning and Management, ASCE*, 123(2), pp. 67-77.

Simpson, A.R., Dandy, G.C. and Murphy, L. J. (1994). "Genetic Algorithms Compared to Other Techniques for Pipe Optimization." *Journal of Water Resources Planning and Management, ASCE*, 120(4), pp. 423-443.

Todini, E. (1999). "A Unifying View on the Different Looped Pipe Network Analysis Algorithms." In R. Powell and K. S. Hindi (Editors) - *Computing and Control for the Water Industry*, Research Press Limited, pp. 63-80.

Todini. E. and Pilati, S. (1987). "A Gradient Method for the Analysis of Pipe Networks." *Proc., Int. Conf. on Comp. Applications for Water Supply and Distribution*, Leicester Polytechnic: Leicester, U.K.

Vairavamoorthy, K. and Ali, M. (2000). "Optimal Design of Water Distribution Systems using Genetic Algorithms." *Computer-Aided Civil and Infrastructure Engineering*, Blackwell Publishers, 15(2), pp. 374-382.

___________. (2005). "Pipe Index Vector: A Method to Improve Genetic-Algorithm-Based Pipe Optimization." *Journal of Hydraulic Engineering*, ASCE, 131(12), pp. 1117-1125.

Wu, Z.Y. and Simpson, A.R. (2001). "Competent Genetic-Evolutionary Optimization of Water Distribution Systems." *Journal of Computing in Civil Engineering*, ASCE, 15(2), pp. 89-101.

Wu, Z.Y. and Walski, T. (2005). "Self-adaptive Penalty Approach Compared with other Constraint-Handling Techniques for Pipe Optimization." *Journal of Water Resources Planning and Management*, ASCE, 131(3), pp. 181-192.

Yates, D.F., Templeman, A.B., and Boffey, T.B. (1984). "The Computational Complexity of the Problem of Determining Least Capital Cost Designs for Water Supply Networks." *Engineering Optimization*, 7(2), pp. 142-155.

Application of Cellular Automata in Civil Engineering

P.J. Gundaliya[1]

ABSTRACT

Earlier computational approaches could model and precisely analyse only relatively simple systems. More complex systems arising in biology, medicine, the humanities, management sciences, and similar fields often remained intractable to conventional, mathematical and analytical methods. Soft computing deals with imprecision, uncertainty, partial truth, and approximation to achieve practicability, robustness and low solution cost and time. In the present study computation technique Cellular Automata is described. Application and use of these soft computing techniques in civil engineering also discussed. Many research is done in recent past using these techniques and found very useful in real-life problems.

Keywords: *Soft computing techniques, cellular automata and heterogeneous traffic flow modelling, routing and scheduling*

INTRODUCTION

The main advantage was an efficient and fast performance when used in computer simulations, due to their rather low accuracy on a microscopic scale. These so-called traffic cellular automata (TCA) are dynamical systems that are discrete in nature, in the sense that time advances with discrete steps and space is coarse grained (e.g., the road is discretized into cells of 7.5m wide, each cell being empty or containing a vehicle). A cellular automata is an n-dimensional array of simple cells where each cell may be in any one of k-states. At each tick of the clock a cell will change its state based on the states of the cells in a local neighbourhood. Typically, the rule for updating the state does not change over time, and is applied to the whole grid simultaneously. Due to its simplicity the CA rules are used to solve the complex behaviour. Through the use of powerful

1. Professor, Civil Engineering Department, L.D. College of Engineering, Ahmedabad – 380 015, Email: pjgundaliya@gmail.com

computers, these models can encapsulate the complexity of the real world traffic behaviour and produces clear physical patterns that are similar to those we see in everyday life. One more advantage of cellular automata models is their efficiency in showing the clear transition from the moving traffic to jamming traffic. CA models have the distinction of being able to capture micro-level dynamics and relate these to macro level traffic flow behaviour.

BACKGROUND

Despite their very simple construction, nothing like general cellular automata appear to have been considered before about the 1950s. Yet in the 1950s – inspired in various ways by the advent of electronics computers – several different kinds of systems equivalent to cellular automata were independently introduced. A variety of precursors can be identified. Operations on sequences of digits had been used since antiquity in doing arithmetic. Finite difference approximations to differential equations began to emerge in the early 1900s and were fairly well known by the 1930s. And Turing machines invented in 1936 were based on thinking about arbitrary operations on sequences of discrete elements.

The best-known way in which cellular automata were introduced (and which eventually led to their name) was through the work by Neumann (1959). The first, mostly in the 1960s, was increasingly whimsical discussion of building actual self-reproducing automata often in the form of spacecraft. The second was an attempt to capture more of the essence of self-reproduction by mathematical studies of detailed properties of cellular automata.

COMPONENTS OF CELLULAR AUTOMATA

There are four components, which play a major role in cellular automata.

The Physical Environment

The term physical environment indicates the physical platform on which CA is computed. It normally consists of discrete lattice of cells with rectangular,

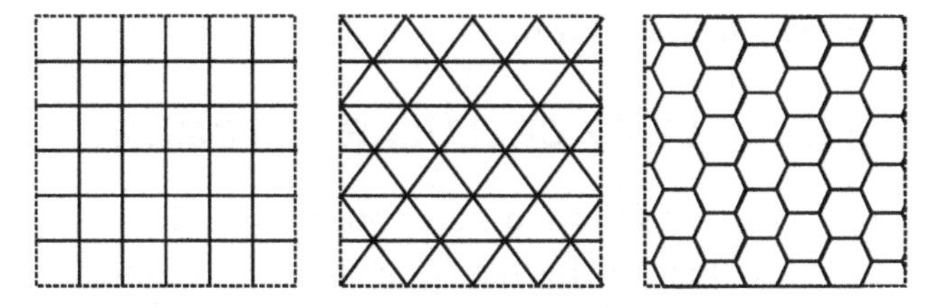

Fig. 1. Physical environment as different types of cells

hexagonal, etc. All these cells are equal in size. They can be finite or infinite in size and its dimensionality can be 1 (a linear string of cells called an elementary cellular automaton or ECA).

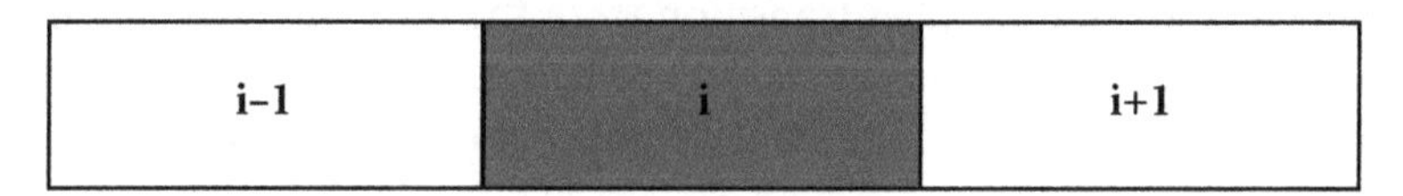

Fig. 2. Neighbourhoods of the present cell i

The Cells States

Every cell can be in a particular state where typically an integer can determine the number of distinct states a cell can be in, e.g. (binary state). Generally, the cell is assigned with an integer value or a null value based upon its state. The states of cells collectively are called as "Global configuration". This convention clearly indicates that states are local and refer to cells, while a configuration is global and refers to the whole lattice.

The Cells' Neighbourhoods

The future state of a cell is mainly dependent on its state of its neighbourhood cell, so neighbourhood cell determines the evolution of the cell. So generally, the lattices vary as one-dimensional and two-dimensional. In one-dimensional lattice, the present cell and the two adjacent cells form its neighbourhoods, whereas in the context of two-dimensional lattice there are four adjacent cells which acts as the neighbourhoods. Therefore, it is clear that as the dimensionality increases the number of adjacent cells also increases.

A Local Transition Rule

This rule (also called function) acts upon a cell and its direct neighbourhood, such that the cell's state changes from one discrete time step to another (i.e., the system's iterations). The CA evolves in time and space as the rule is subsequently applied to all the cells in parallel. Typically, the same rule is used for all the cells (if the converse is true, then the term hybrid CA is used). When there are no stochastic components present in this rule, we call the model a deterministic CA, as opposed to a stochastic (also called probabilistic) CA.

WORKING PRINCIPLE OF CELLULAR AUTOMATA

In cellular automata space, time and state variables are discrete which makes them ideally suited for high-performance computer simulation. However, CA modelling differs in several respects from coutinum models. These are usually based on differential equations, which often cannot be treated analytically. One has to solve them numerically and therefore the equations have to be discretized. In general, only space and time variables become discrete whereas the state variable (e.g. the density or velocity) is still continuous. CA is discrete in space and time variables, this discreteness are already taken into account in the definition of the model and its dynamics. This allows for obtaining the desired behaviour in a much simpler way. The numerical solution of (discretized) differential equations is only accurate in the limit Δx, $\Delta t \rightarrow 0$. This is different in the CA where Δx and Δt are finite and accurate results can be obtained since the rules (dynamics) am designed such that the discreteness is an important part of the model.

In order to achieve complex behaviour in a simple fashion one often resorts to a stochastic description. A realistic situation seldom can be described completely by a deterministic approach. A cellular automaton is a discrete dynamical system. Each point in a regular spatial lattice, called a cell, can have anyone of a finite number of states. The states in the cells of a lattice are updated according to a local rule say R. That is, the state of the cell at a given time depends only on its own state one time step previously, and the states of its nearby neighbours at the previous time step. All cells in the lattice are updated synchronously. The state of the lattice advances in discrete time steps. In the above definition, the rule R is identical and homogeneous for all sites and applied simultaneously to each of them. The rule R is some well-defined function and a given initial configuration will always evolve the same way.

However it may be very convenient for some applications to have certain degree of randomness in the rule. It may be desirable for some instance, that a rule selects one outcome among the several possible states; with a probability p. Cellular automata whose updating rule is driven by external probabilities are called probabilistic cellular automata. On the other hand, those which strictly comply with the definition given above are referred to as deterministic cellular automata.

CA is used in traffic flow modelling since last decade. The freeway being simulated is discretized into homogeneous cells of equal length, and time is discretized into time-steps of equal duration. These cells can be either in an occupied or empty state, depending on whether a vehicle is present at that location. The state of the cells is updated sequentially at each time step with a set of vehicle position updating rules.

A cellular automata rule is local, by definition. The updating of a given cell requires one to know only the state of the cells in its vicinity. The spatial region in which a cell needs to search is called the neighbourhood. For two-dimensional cellular automata, two neighbourhoods are often considered. The key idea of neighbourhood is that when updation occurs, the cells within the block evolve only according to the state of that block and don't depend on what is in the adjacent blocks. In practice when simulating a given cellular automata rule, it is not possible to deal with an infinite lattice. The system must be finite and have boundaries. A site belonging to the lattice boundary doesn't have the same neighbourhood as other internal sites. In order to define the behaviour of these sites, a different evolution rule can be considered, which sees the appropriate neighbourhood. The basic deterministic CA rules given by Wolfram (1986) is explained in following subsection.

WOLFRAM'S CA RULES

Wolfram (1986) worked with a one-dimensional variant of von Neumann's cellular automata; this was fully horizontal and occurred on a single line. Each cell touched only two other cells, its two immediate neighbours on either side, and each succeeding generation was represented by the line underneath the preceding one. A cell in generation two would determine its state by looking at the cell directly above it, i.e. in generation one, and that cell's two neighbours. Thus, there are eight possible combinations of the states of those three cells ranging from "000" (all off, all white) to "111" (all on, all black) as shown in Fig. 3. Three cells and their updating state called an applet. He has specified such different combination and given them CA rule number from 0 to 254. The

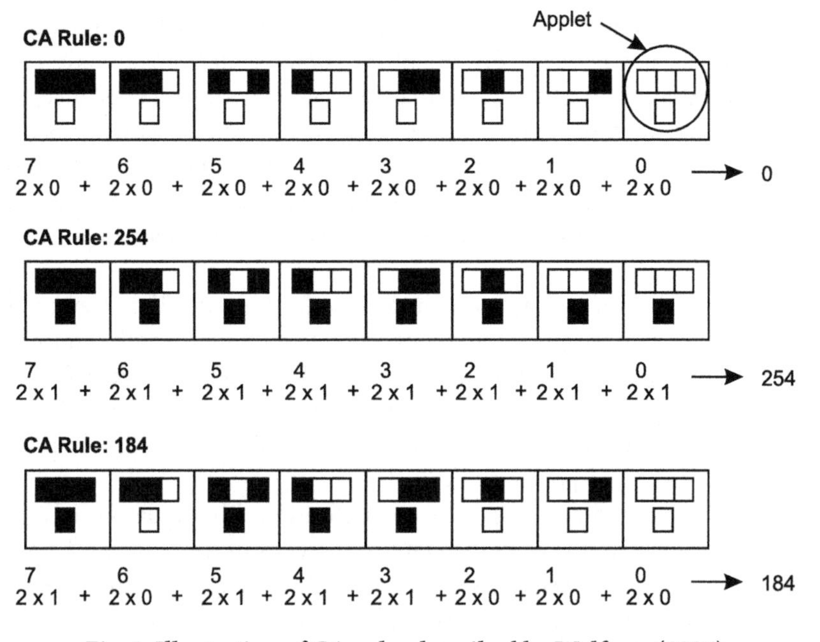

Fig. 3. Illustration of CA rules described by Wolfram (1986)

CA rule 0, CA rule 254 and CA rule 184 are illustrated in Fig. 3. The CA rule 0 means that whatever updation occurs in the neighbouring cells, the middle cell will be white (fixed rule).

Application of CA Rule: 184 in vehicle movement

Cell Numbers	12	11	10	9	8	7	6	5	4	3	2	1
TIME STEP:1				■					■			
TIME STEP:2					■					■		
TIME STEP:3						■					■	

Fig. 4. Updation of cell status in case of CA rule 184

Similarly CA rule 254 updates middle cell to black for any color combination in neighbouring cells. The working of CA rule 184 is illustrated as shown in Fig. 4. In this rule, status of cell updation is defined as described in CA rule 184 in Fig. 4. The cells update their state according to neighbour's state (flexible rule). Hence, in the second time-step the cell number four, five, ten and eleven from the left, changes its state in next time step according to rule 184 as shown in Fig. 4. This is similar to the position updation of vehicles (or how the vehicle advances) at every time step. Cellular automata are the mathematical models for complex natural systems containing large numbers of simple identical components with local interactions. They consist of a lattice of sites, each with a finite set of possible values. The value of the sites evolves synchronously in discrete time steps according to identical rules. The value of a particular site is determined by the previous values of a neighbourhood of sites around it. Single dimension cellular automata are arrays of discrete cells with discrete values. Yet sufficiently large cellular automata often show seemingly continuous macroscopic behaviour. They can thus potentially serve as models for continuum systems, such as fluids. Their underlying discreteness, however, makes them particularly suitable for digital computer simulation and for certain forms of mathematical analysis. On a microscopic level, physical fluids also consist of discrete particles. But on a large scale, they, too, seem continuous, and can be described by the partial differential equations of hydrodynamics. The form of these equations is in fact quite insensitive to microscopic details. Changes in molecular interaction laws can affect parameters such as viscosity, but do not alter the basic form of the macroscopic equations. As a result, the overall behaviour of fluids can be found without accurately reproducing the details of microscopic molecular dynamics.

Cellular automata is developed in discrete dynamics of space and time, and is a discrete simulation method. It is buildup from the single string of one-dimensional automata, and can be arranged in two or higher dimensional lattice for two or higher dimensional automata. All cells in CA are identical and having discrete state. The future state of each cell depends only of the current state of the cell and the states of the cells in the neighbourhood. The development of each cell state is defined by simple rule CA models are in principle amenable to single bit coding. This approach runs extremely fast on traditional vector-computers.

Properties of CA

- CA developed in space and time.
- A CA is a discrete simulation method; hence space and time are defined in discrete steps.
- A CA is built up from cells that are lined up in a string for one-dimensional automata.
- CA can be arranged in a two or higher dimensional lattice for two- or higher dimensional automata
- The number of states of each cell is finite.
- The states of each cell are discrete.
- All cells are identical.
- The future state of each cell depends only of the current state of the cell and the states of the cells in the neighbourhood.
- The development of each cell is defined by rules.
- CA models are in principle amenable to single bit coding. This approach runs extremely fast on traditional vector-computers.

The Advantages of CA Traffic Flow Model

The use of the cellular automata in modelling has certain advantages, as compared to other types of models. The most important advantages are the following:

- The model is simple. And it is easy to be achieved on the computer.
- It can re-create all kinds of complicated traffic phenomena and reflect the properties of traffic flow.
- The roads are divided into plenty of minute lattices. The conditions of cars turning are simplified as travelling in straight lines. It can simplify the parameters of roads.
- CA model is a dynamic model, which consists of limitless discrete of space, limited discrete of state, all discrete time is integer.

- Once the local interactions between the cells are solved, the system can be increased up to any size, without any other modelling problems.
- The results of the scientific research and the experiments reported in the scientific literature prove that the traffic simulation with cellular automata is an interesting and useful research topic.

Limitations of CA for Traffic Flow Modelling

Some of the limitations of CA based traffic flow models are listed below.

- Time setup is discrete hence the acceleration and deceleration is more than the real.
- If vehicle size and speed are widely differ, then it's difficult to represent them as uniform cells.
- Accelerations and decelerations are much larger than in reality due to discreteness.
- Lane change is done in one second (or for given time-step) whereas actual time required more than that.
- The speed of vehicle is imitated to discrete time steps.

The traffic flow modelling using CA is extensively done for uninterrupted roads like freeway and arterial to model the traffic flow behaviour. In cellular automata the cells, which are either empty or occupied by exactly one vehicle. Movement takes place by distance between vehicle and vehicles characteristics. In CA models, a road is represented as a string. Initial proposition of a CA model for traffic is given by Gerlough (1956) and same is extended by Cremer and Papageorgio (1981), Cremer and Ludwig (1986) and co-workers. They implemented fairly sophisticated driving rules and also used single-bit coding with the goal to make the simulation fast enough to be useful for real-time traffic applications. The bit-coded implementation, though, made it too impractical for many traffic applications. In 1992, CA models for traffic were brought into the statistical physics community. Nagel and co-workers used model with maximum velocity for one-and for two-dimensional traffic. One-dimensional here refers roads includes multi-lane traffic. Two-dimensional traffic in the CA context usually means traffic on a two dimensional grid, as a model for traffic in urban areas (intersection).

Cell Size

Deciding the cell size is crucial for the grid-based simulation approach as it affects computational efficiency and performance of the model. In the present study, a systematic approach is adopted to decide the cell size in cellular automata based modelling for heterogeneous traffic flow. In this model, size of the cell is carefully decided according to the type of vehicle. It is decided in such a way that it represents the actual size of vehicles and the total width of the road as close as possible. The physical representation of the vehicle should be kept slightly more than the actual size of vehicle to provide some clearance. The cell length also depends upon the dynamic characteristics of the vehicular movement, as in the cellular automata, distance-headway and speed is considered in terms of number of cells.

NAGEL-SCHRECKENBERG (NASCH) MODEL

Nagel and Schreckenberg (1992) presented the cellular automata model for single lane traffic flow for homogeneous traffic. This model is taken as the base model for present research work. The process of vehicle movement is described here. In CA traffic flow models, the position, speed, acceleration as well as time are treated as discrete variables. In a basic CA traffic flow model each vehicle has an integer speed with values between zero and the maximum speed of the vehicle. The speed of each vehicle can take one of the integer values out of say 0, 1, 2, 3, ... , V_{max}, in term of cells per time step. The road is represented by a cell which can be either empty or occupied by at most one vehicle at a given instant of time. At each discrete time step, the state of the system is updated following a well-defined rule. In NaSch model vehicles updation are followed by four rules namely acceleration, deceleration, randomization and updation. This model is developed for the homogeneous traffic flow. In these rules except probability p all operation are integer. A numerical example is taken for explaining vehicle movement on road in NaSch model. In this example noise probability (*p*) taken as 1/3 which reflect the driver behaviour. All vehicles are taken as cars with maximum speed (V_{max}) allowed as 2 cell/time step. In Fig. 5 the road stretch are represented with the uniform front gap of vehicle is taken numerically as number of empty cells ahead plus one. The updation of vehicle speed is indicated after applying each rule shown in Fig. 5. The uniform vehicle (car) is considered in this example. Due to noise probability an average one third of the cars qualified slowdown in the randomization step. All cars update their speed parallely at every time step as shown in Fig. 5 step by step. In Fig. 5 all four cars speed are indicated on right top corner of the cell at every step. The initial speed of car 1, 2, 3 and 4 are 0, 1, 2, and 1 respectively. The front gap of all cars 1, 2, 3, and 4 are -1, 1, 3, and 2 respectively. The front gap of all cars is indicated at the right bottom corner of the cell in Fig. 5. The first car front gap is taken as -1, means that there is infinite gap and vehicle can attain the speed of its maximum speed.

Step 1: Acceleration

If $V_n < V_{max}$ then the speed of the nth vehicle is increased by one, $V_n = min\ (V_n + 1, V_{max})$ but remains unaltered if $V_n = V_{max}$. After applying this rule the speed of the cars are 1, 2, 2, and 2 respectively for car 1, 2, 3, and 4. The car 3 is going with its maximum speed hence its speed remain unaltered. This reflects the general tendency of the drivers to drive as fast as possible, if allowed to do so, without crossing the maximum speed limit.

Step 2: Deceleration due to Other Vehicles (vehicle ahead)

If $gap_p^f \leq v_n$ in the speed of the nth vehicle is reduced to $gap_p^f - 1$, here gap is one even if a front vehicle is there in the next cell, $v_n = max\ (v_n, gap_p^f - 1)$. Here gap_p^f is front vehicle gap in the present lane. After applying this rule the speed of cars are 1, 0, 2, and 1 respectively for car 1, 2, 3, and 4. The car 2 and car 4 reduces its speed based on the available front gap in number of cell as 0 and 1 cells per time step respectively.

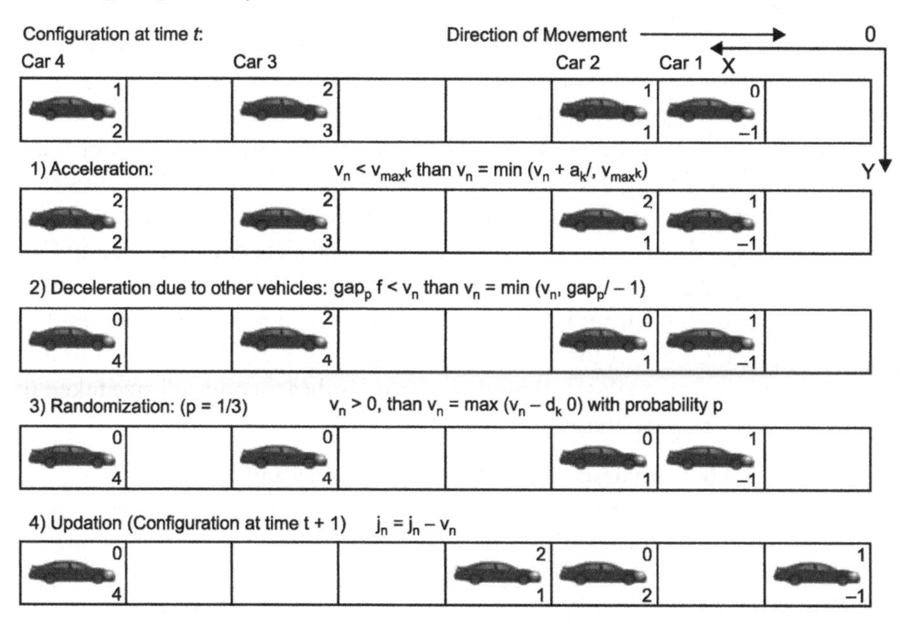

Fig. 5. Numerical example for the application of CA rules for NaSch model

Step 3: Randomization

If $v_n > 0$, the speed of the nth vehicle is decreased randomly by one unit with the probability p but does not change $v_n = 0$, $v_n = max\ (v_n - 1, 0)$ but remains unaltered if $v_n = v_{max}$. After applying this rule the speed of the cars are 1, 0, 2, and 0 respectively for car 1, 2, 3, and 4. After applying this rule the car 4 has its speed reduced by 1 with the chance of probability p. This step takes into account the different behavioural patterns of the individual drivers, especially, non-deterministic acceleration and overreaction while slowing down; this is crucially important for the spontaneous formation of traffic jams.

Step 4: Vehicle Movement

Each vehicle moves forward according to its new speed determined in Steps 1-3, i.e.

$$x_n = x_n + v_n$$

The vehicle then forwarded with the number of cell (speed) as shown in step 4 in Fig. 3. After applying all rule the new speed of cars are 1, 0, 2, and 0 respectively for car 1, 2, 3, and 4. The front gap all cars are then updated as -1, 2, 1, and 4 respectively for car 1, 2, 3 and 4. This procedure again repeated for another time step. It is to be noted that even changing the precise order of the steps of the update rule stated above would change the properties of the model. This model may be regarded as a stochastic CA.

The literature shows the feasible direction of research in the area of traffic flow model. Cellular automata found a new computational effective way to simulate the heterogeneous traffic. CA has proved its ability to predict the behaviour of traffic phenomenon and overcome some of the disadvantages of the mathematical modelling. Due to the discrete nature of CA traffic flow model, it becomes more prevalent for computer implementation and simulation. Many researchers have developed homogeneous traffic models effectively using CA. They have showed that the microscopic behaviour of the traffic flow can be reasonably predicted through CA models. However, many efforts required to explore application of CA for the heterogeneous traffic. So there is a possibility for development and analysis. Table 1 contains study area and cell size taken in model for study in past literature.

Table 1. Analysis of literature on study area and cell-size

Author name	Study area	Cell size
Lan at al (2005)	Provincial Highway	1.25 m (Width) × 1.25 m (length)
Gundaliya et al (2006)	Urban Arterial road	3.6 m (Width) × 5 m (length)
Gundaliya et al (2008)	Urban Arterial road	0.9 m (Width) × 1.9 m (length)
Mallikarjuna et al (2009)	Urban mid-block	1.4 m (Width) × 0.5 m (length)
Pal et al (2010)	Urban mid-block section	Depend on dominating vehicle
Meng et al (2011)	Urban work-zone area	0.7 m (Width) × 0.5 m (length)
Mu et al (2012)	highway and an arterial road	4 m (length)
Radhakrishnan et al (2013)	Urban signalized intersection	1 m (Width) × 1 m (length)

Application

In a cellular automata based Mid-Block Traffic Flow Model (MBTFM) is developed, the road space is divided into imaginary cells of equal size. The forward movement of the vehicle is implemented using CA rules similar to the Nagel and Schreckenberg (1992) model known as NaSch model unlike other grid based models. Rules for the free and forced lateral movement have also been proposed in the study. The complete methodology of the developed model is described in flow chart of Fig. 6. The major components of the model are input, model initialization, application of CA rules, vehicle generation, and get desired output.

The required input is cell size, road length in terms of the number of cells for defining the road stretch, the vehicle size in terms of the number of cells (length and width), arrival patterns (distribution of headway, speed), and arrivals lateral position distribution. In addition, the maximum speed (cells per time step), acceleration, and

Deceleration of each type of vehicle, lateral movement probabilities are required as input for the model. For the study of an incident effect (Pedestrian Signal), the model needs location and duration of the incident as input. In the case of providing Pedestrian signal on the urban mid-block road section, the green time, red time and amber time are required to be specified along with the location of signal.

Cell Size

In NaSch models, researchers assumed that each cell has a uniform length of 7.5 m and cell width is equal to lane width 3.6 m. Each vehicle has the uniform size and occupies one cell. Obviously, such cell size is too uneven, which leads to unrealistic acceleration and deceleration rates. The uniform vehicle size cannot account for the fact that different types of vehicles have different vehicle sizes.

Cell size affects computational efficiency and performance of the model. In the present study, a systematic approach is adopted to decide the cell size in cellular automata based modelling for heterogeneous traffic flow. In this model, size of the cell is carefully decided according to the type of vehicle. The physical representation of the vehicle should be kept slightly more than the actual size of vehicle to provide some clearance. The cell length also depends upon the dynamic characteristics of the vehicular movement, as in the cellular automata, distance-headway and speed are considered in terms of number of cells.

A minimization problem is formulated considering headway distance, total number of cells required to represent vehicle types, and road width. The first term in the objective function represents the headway difference such that

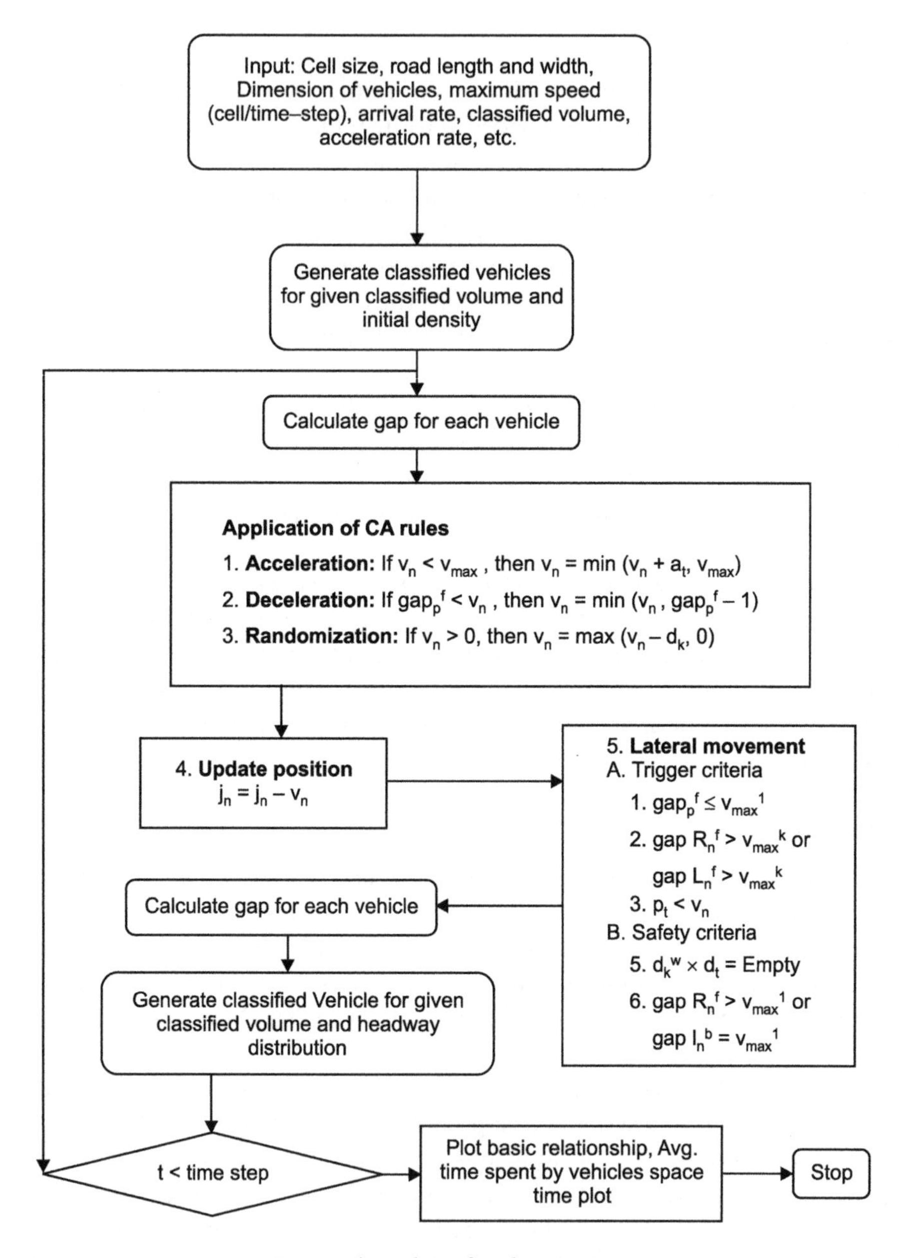

Fig. 6. Flow chart for the MBTFM

the model reacts similar to the NaSch basic model at all the cells with six speeds options. According to CA, the minimum headway required is the number of cells ahead of the vehicles as per the vehicle speed. Speeds and headways for

NaSch model compared with MBTFM model are shown in Table 2. In this study takes cell size is 0.9 m × 0.9 m.

The difference between headways of NaSch model and GBTFM is compared with that of MBTFM model at the same speed. Minimum difference indicates that the model closely represents the dynamic characteristics of NaSch model in metres.

Table 2. Speed and distance headway for MBTFM models

Speed and distance headway for CA models				
Speed in (Cell/Time-step)	NaSch model (3.6 m × 7.5 m)		MBTFM (0.9 m × 0.9 m)	
	Speed (Km/h)	Headway (m)	Speed (Km/h)	Headway (m)
3	81	22.5	9.72	2.7
4	108	30	12.96	3.6
5	135	37.5	16.2	4.5
6	-	-	19.44	5.4
7	-	-	22.68	6.3
8	-	-	25.92	7.2
9	-	-	29.16	8.1
10	-	-	32.4	9
11	-	-	35.64	9.9
12	-	-	38.88	10.8
13	-	-	42.12	11.7
14	-	-	45.36	12.6
15	-	-	48.6	13.5
16	-	-	51.84	14.4
17	-	-	55.08	15.3
18	-	-	58.32	16.2
19	-	-	61.56	17.1
20	-	-	64.8	18
21	-	-	68.04	18.9
22	-	-	71.28	19.8
23	-	-	74.52	20.7

(Contd.)

(Table 2 *contd.*)

24	-	-	77.76	21.6
25	-	-	81	22.5
26	-	-	84.24	23.4
27	-	-	87.48	24.3
28	-	-	90.72	25.2
29	-	-	93.96	26.1
30	-	-	97.2	27
31	-	-	100.44	27.9
32	-	-	103.68	28.8
33	-	-	106.92	29.7
34	-	-	110.16	30.6
35	-	-	113.4	31.5
36	-	-	116.64	32.4
37	-	-	119.88	33.3
38	-	-	123.12	34.2
39	-	-	126.36	35.1
40	-	-	129.6	36
41	-	-	132.84	36.9
42	-	-	136.08	37.8

Vehicle Size

After deciding cell size, length and width of the vehicle in number of cells can be obtained by adding clearance to the vehicle's actual length and width. After deciding the size of the vehicle in terms of number of cells in lateral and longitudinal directions, vehicle can be physically represented on occupied number of cells. The left most corner of each vehicle represents the position of the vehicle in each time step.

Seven different vehicle types are simulated: two wheeler (2W), three wheeler (3W), Passenger-car, LCV1, LCV2, HCV1 and HCV2. (Fig. 7)

Table 3 shows vehicle type, the actual dimensions of width and length of vehicles in metres, width and length of vehicle taken in present model in metres, shows its dimensions of width and length in cells taken in model respectively, and shows the minimum clearance on width and length.

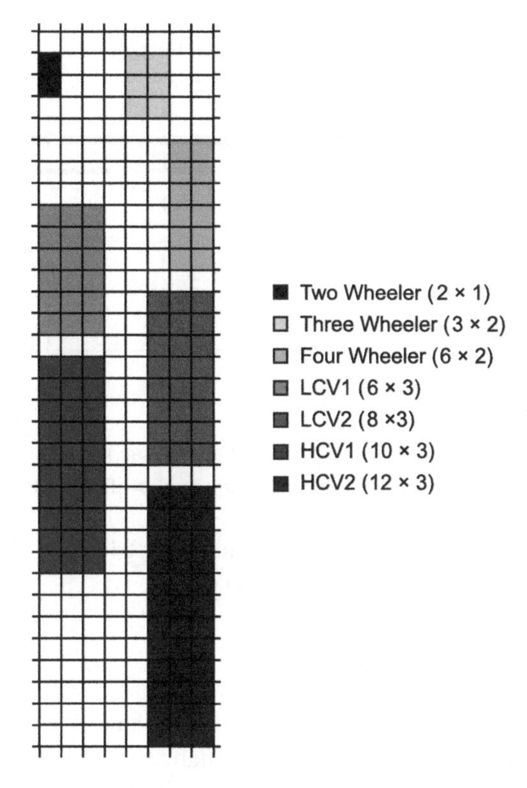

Fig. 7. Each vehicle shows with occupied cells (cell size 0.9 m × 0.9 m)

Table 3. Vehicles dimensions and clearance

Vehicles dimension and clearance for cell size 0.9 m × 0.9 m									
Sr. No.	Vehicle Type	Actual		Taken in model		Taken in model		Clearance	
		Width (m)	Length (m)	Width (m)	Length (m)	Width (Cells)	Length (Cells)	Width (m)	Length (m)
1.	2W	0.6	1.8	0.9	1.8	1	2	0.3	0
2.	3W	1.4	2.6	1.8	2.7	2	3	0.4	0.1
3.	Car	1.7	4.7	1.8	5.4	2	6	0.1	0.7
4.	LCV1	1.9	5	2.7	5.4	3	6	0.8	0.4
5.	LCV2	2.2	6.8	2.7	7.2	3	8	0.5	0.4
6.	HCV1	2.5	8.5	2.7	9	3	10	0.2	0.5
7.	HCV2	2.5	10.3	2.7	10.8	3	12	0.2	0.5

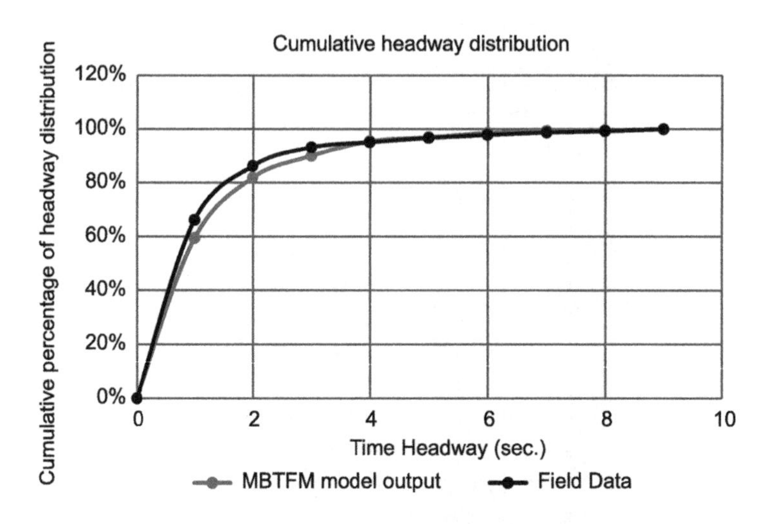

Fig. 8. Cumulative percentage plot for headway distribution of field data and MBTFM model output

MACROSCOPIC VALIDATION

To investigate the model validity, we also perform a macroscopic evaluation to identify the overall MBTFM model performance. In macroscopic validation, the speed-flow-density relationship is investigated to evaluate how well the MBTFM model performs.

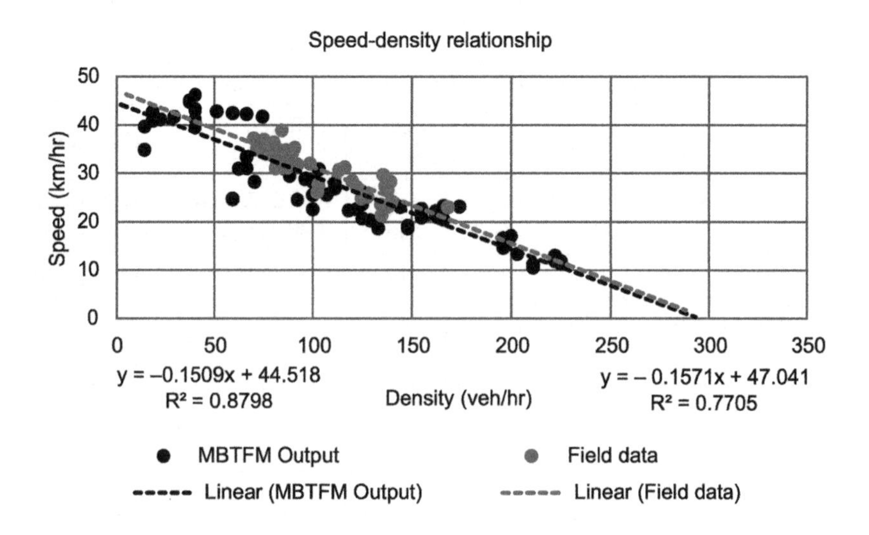

Fig. 9. Speed-density relationship

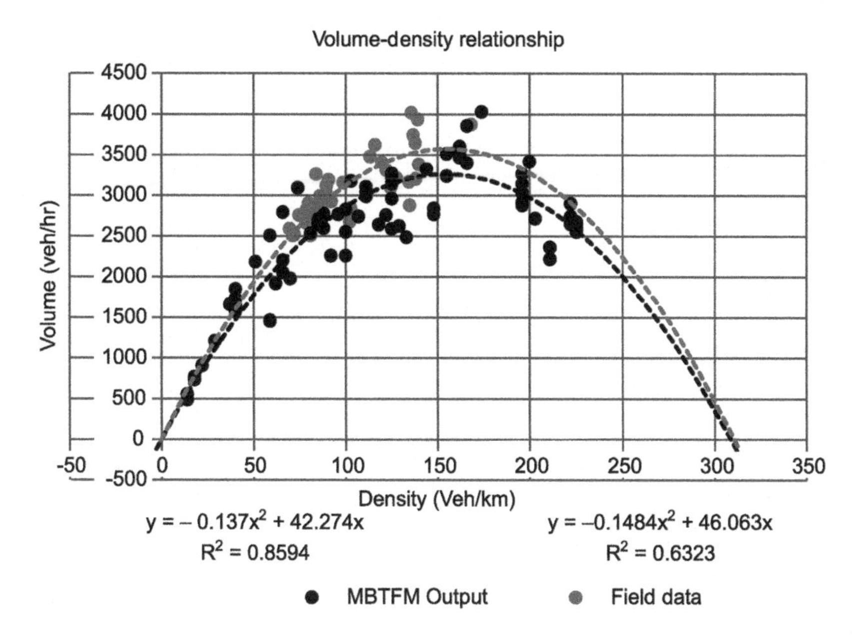

Fig. 10. Volume-density relationship

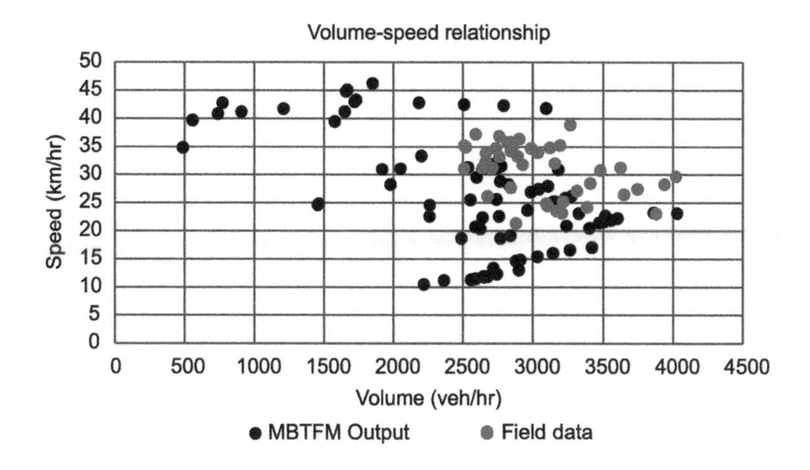

Fig. 11. Volume-speed relationship

Fundamental Relationships from Simulation Experiments

MBTFM can be applied to derive fundamental relationship of different traffic flow parameters in various traffic conditions. In order to investigate the capability of the model to simulate various traffic phenomena, the model is applied to 500.4 m and 7.2 m wide two lane road stretch of road for one hour duration.

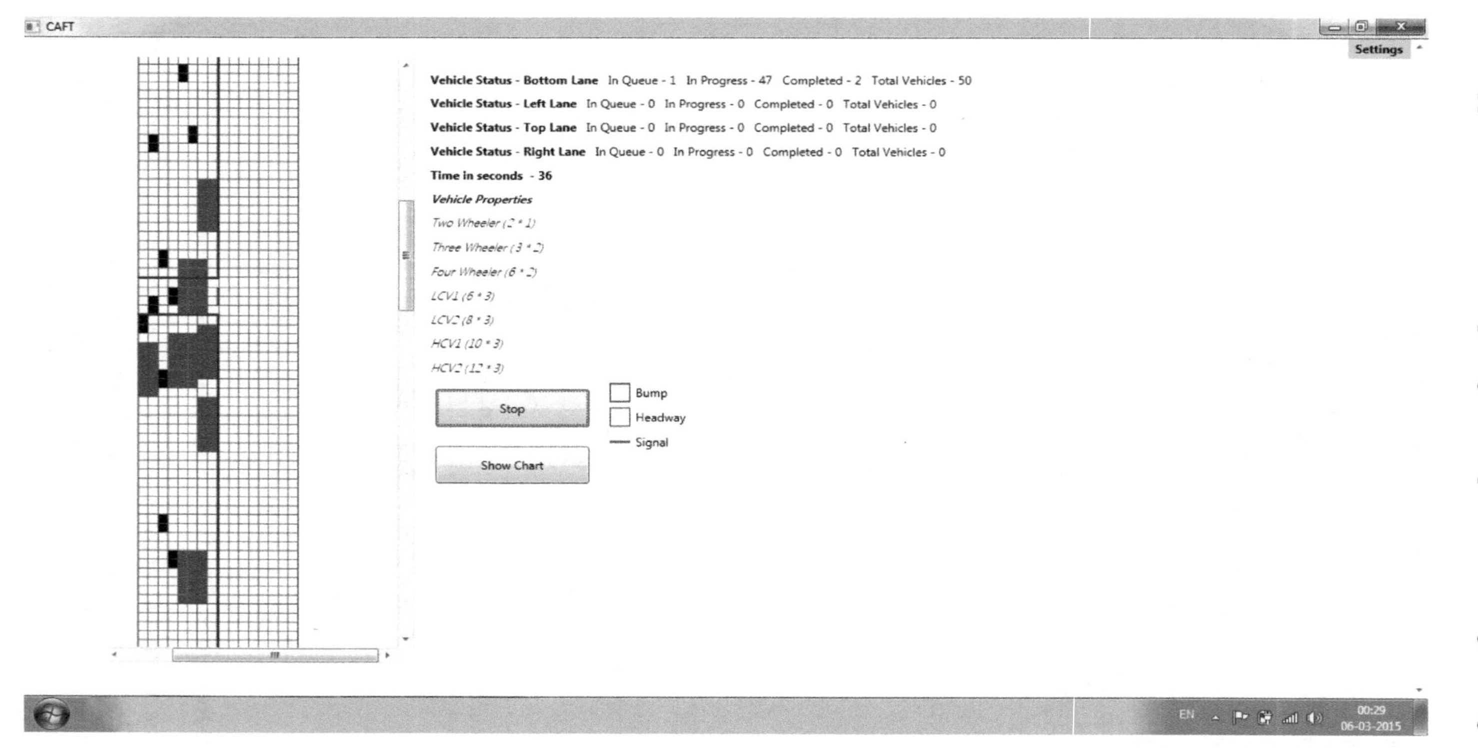

Fig. 12. Screen-shot MBTFM model

REFERENCES

Ahmedabad Bus Rapid Transit System (2006), Working Paper, Centre for Environmental Planning and Technology University, Ahmedabad.

Ahmedabad Bus Rapid Transit System (2006), Working Paper, CEPT, Ahmedabad.

Alam, M. Absar and Ahmed, Faisal (2013). Urban Transport Systems and Congestion – A Case Study of Indian Cities, *Transport and Communications Bulletin for Asia and the Pacific*, 82, pp. 33-43.

BRTS -DPR, Phase-I , February-2007 AUDA , AMC and Centre for Environmental Planning and Technology (CEPT) University, Ahmedabad.

BRTS–DPR, Phase-II, April-2008. Centre for Environmental Planning and Technology University, Ahmedabad.

Dinesh Mohan. Public Transportation Systems for Urban Areas – A Brief Review.

Ganti, R. (2001). *Study of Overtaking Manoeuvres in Mixed Traffic Flow Using Computer Simulation.* M.Tech. thesis (unpublished), I.I.T. Madras.

Gundaliya, P.J., Mathew, Tom V. and Dhingra, S.L. (2005). "Heterogeneous Traffic Flow Modelling using Cellular Automata", World Transport Research - *Proceedings from the 10th World Conference on Transport Research*, Istanbul, Turkey.

Khan, S.I. and Maini, P. (2000). "Modelling Heterogeneous Traffic Flow." *Transportation Research Board,* 1678, pp. 234-241.

Kharola, P.S., Tiwari, Geetam and Mohan, Dinesh (2010). "Traffic Safety and City Public Transport System: Case Study of Bengaluru, India. *Journal of Public Transportation,* 13(4), pp. 63-91.

Korlapati, D.R. (2003). *Evaluation of Diversion Strategies for an Urban Traffic Corridor with Heterogeneous Traffic.* Ph. D. thesis, Indian Institute of Technology, Madras (India).

Koshy, R.Z. and Arasan, V.T. (2005). "Methodology for modeling highly heterogeneous traffic flow." *Journal of Transportation Engineering, ASCE,* 131(7), pp. 544-551.

Marwah, B. and Ramaseshan, S., (1978). Interaction between Vehicles in Mixed Traffic Flow using Simulation, Indian Roads Congress, *Highway Research Bulletin*, 8, pp. 1-15.

May, A.D. (1990). *Fundamentals of Traffic Flow.* Prentice-Hall, Inc.: Englewood Cliffs, New Jersey, 07632, Second edition.

Ministry of Road Transport and Highway Outcome Budget – 2013-14, Government of India.

National Transport Development Policy Committee (NTDPC) (2014), India Transport Report: Moving to 2032.

O.P. Agarwal. Urban Transport in India.

Pucher, John, Korattyswaroopam, Nisha and Ittyerah, Neenu. (2004). "The Crisis of Public Transport in India: Overwhelming Needs but Limited Resources". *Journal of Public Transportation*, 7(4), pp. 1-20.

Rajagopal, A. and Dhingra, S.L. (2002). "Simulation-Based Evaluation for Traffic Management." *Proc., National Conference on Transportation Systems (NCTS)*, Indian Institute of Technology, New Delhi (India), pp. 449-458.

Ramanayya, T.V. (1980). *Simulation Studies on Traffic Capacity of Road Systems for Indian Conditions. Ph.D. thesis,* University of Kakatiya.

___________. (1988). Highway Capacity under Mixed Traffic Conditions, *Traffic Engineering and Control,* pp. 284-287.

Road Transport Annual Report 2011-12, Ministry of Roads Transport and Highway, Government of India.

Sanjeev Kumar Lohia. Urban Transport in India.

10

Optimization of Cropping Pattern in Karjan Command Area Using Linear Programming

S.M. Yadav

INTRODUCTION

The history of irrigation development in India can be traced back to prehistoric times. Vedas and ancient Indian scriptures made references to wells, canals, tanks and dams which were beneficial to the community and their efficient operation and maintenance was the responsibility of the State. Civilization flourished on the banks of the rivers and harnessed the water for sustenance of life. In a monsoon climate and an agrarian economy like India, irrigation has played a major role in the production process. There is evidence of the practice of irrigation since the establishment of settled agriculture during the Indus Valley Civilization (2500 BC).

The need to conserve and manage water is not new, although in the past, most countries could meet their requirements for domestic, industrial and agricultural supplies well within the water resources available to them. As population increases with time, demands per capita for domestic and industrial consumption also increases because of improved standards of living. The situation is most acute in countries like India which are heavily dependent on irrigation to meet their domestic food needs. This impeding crisis is unlikely to be recognized as an absolute resource constraint unless planning methods are adopted which are designed to investigate water resources in a comprehensive manner. Therefore there is a felt of need for a fundamental reassessment of the methods of analysis and water management adopted and the development of a new framework and new techniques. Well, during the last 20 years, one of the most important advances made in the field of water resources engineering is the development and adoption of optimization techniques for planning, design and management of complex water resource systems. Extensive Literature review of the subject reveals that many successful applications of these techniques have been made.

A number of reservoirs have been planned and constructed in India for conservation utilization of the water resources for deriving various benefits including flood control. In Indian context, about 80% of its population mainly depends on agriculture.

Under the present situation of rapid population growth, there is always a heavy pressure for exploitation of available water resource in the best possible manner to meet the increasing demand. The total geographical area in India is 329 mha of which 47% is cultivated area which accounts to about 154 mha and the rest 53% comes under non-cultivable , barren and forested area. 37% of cultivated area i.e. 56.98 mha is the irrigated area and produces 55% of net total revenue whereas 63% of cultivated area i.e. 97.02 mha produces only 45% of net revenue. The projected demand for irrigation for the year 2010 is 78% of the total demand and is predicted to be 72% in the year 2025. These statistics show the vital importance of irrigation in our country for which reservoir operation plays a major role.

There is ever increasing need for irrigation for better agricultural outputs. Under such circumstances, it is essential to manage available water and its resources by the concerned water resources engineers or hydrologists in the best possible manner. Hence, optimization of irrigation and reservoir operation is an area to be focused upon.

NEED FOR STUDY

To optimize his decision, the farmer must choose among the available production alternatives, the most efficient in the use of productive resources and the one which satisfies the previously-stated goals. In the cases where the decision is related to the allocation of scarce resources, the farmer's responsibility is to find efficient methods that can help him to make the right decision.

To solve this problem, the mathematical programming models are the most recommended. The mathematical programming quantifies an optimal way of combining scarce resources to satisfy the proposed goals. That is why, the cases, where the available resources must be combined in a way to maximize the profit or minimize the cost should be analysed.

OBJECTIVE OF STUDY

The main objective of this study is to obtain optimum cropping pattern, net benefits and evaporation losses, storage and over flow on monthly basis considering 75% dependable flow and to compare them with actual values.

SCOPE OF THE STUDY

The scope of the study is as follows:

- To develop inflow-outflow model and compute various dependability inflows using mass curve method.
- To develop linear regression model for evaporation.
- The linear programming model for obtaining maximum net benefits of Karjan irrigation scheme is developed.
- To compute crop area, reservoir releases for irrigation, initial reservoir storages, reservoir evaporation and overflow for 75% dependable and compare with the actual values.

DESCRIPTION OF RIVER

The river Karjan, one of the major left bank tributaries of the Narmada, is an interior state river running through Madhya Pradesh, on north boundary of Maharashtra state and through Gujarat state. It emerges out of the Satpura hill ranges near village Kalvipura of Mangrol taluka of Surat district (Gujarat) at altitude of about 350.61 m. It flows and meets the Narmada river about 19°31°NNW of Rajpipla. The length of the river from the above confluence to Narmada river is 32.20 km.

The total length of river from its outh is 96.55 km. The Karjan dam is situated on it at 64.35 km.

Data Collection

The data collected for the study is as listed below:

- The data of inflow in the river Karjan is collected from the year 1994 to 2010.
- The data of storage capacity, evaporation losses from the year 1994 to 2011 (18 years) is collected for the reservoir Karjan.
- The data of overflow from the year 1994 to 2011(18 years) is collected for the dam Karjan.
- The data of actual cropping pattern in hectare of C.C.A. of RBMC (sec 1 to 4) and LBMC (sec 5 to 16) is collected from the year 1999 to 2010 (for 11 years).
- The data for canal carrying capacity is collected.
- The data of net irrigation water requirement and charges of surface water is collected for each crop.

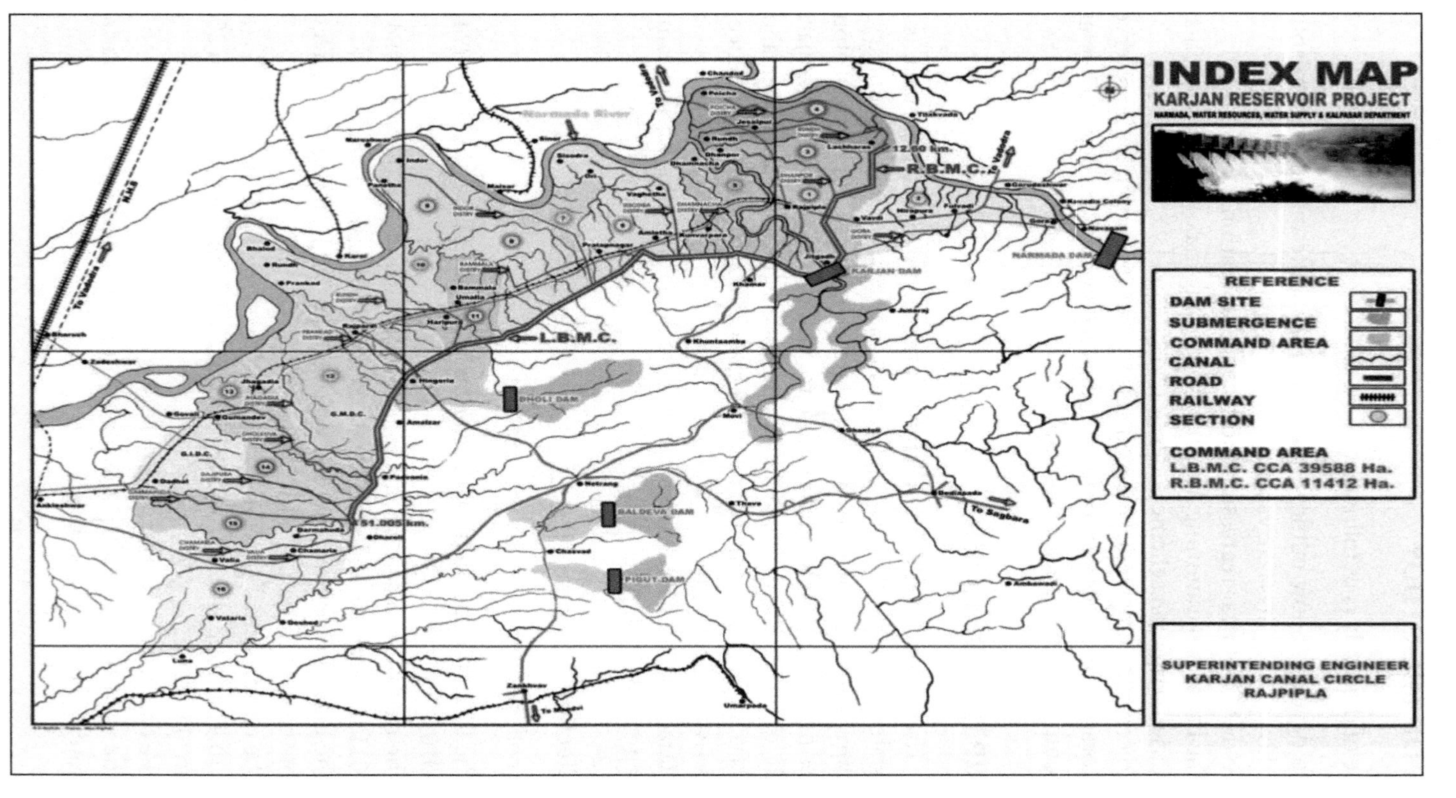

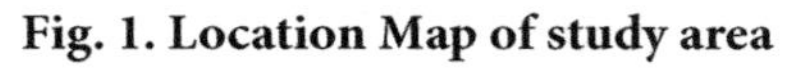
Fig. 1. Location Map of study area

Methodology

A linear programming (LP) model has been developed to derive the optimal pattern for the Karjan irrigation scheme.

The basic steps followed while formulating the objective function and constraints are as follows:

1. Defining the objective function according to the problem (min or max).
2. Define the decision variables.
3. Mathematical formulation of the objective function.
4. Description of corresponding constraints.
5. Mathematical formulation of constraints.

The LP model is prepared for three different considerations.

- LP model-1 is prepared at 75% dependable inflow with maximum crop area sown in past 5 years.
- LP model-2 is prepared at 75% dependable inflow with planned crop area sown.
- LP model-3 is prepared at 75% dependable inflow with average crop area sown in past 5 years.

Formulation of Model

The linear programming model formulation consists of objective function and various constraints for which maximization or minimization is done. In this section various steps involved in the formulation of model has been discussed.

Objective Function

The objective function of the model is to maximize the net benefit accrued over a year in two seasons and is mathematically represented as

$$\text{Max NB} = \Sigma^{9}{}_{i=1}\, C_i ak_i + \Sigma^{9}{}_{i=1}\, C_i bk_i$$

Where

NB = Net benefit accrued in the system in Rs.

i = index of crop, n = 1, 2, 3... 9

C_i = cost coefficient of crop "i" assuming full irrigation proposed on deducting the fertilizer and labour cost from the gross benefits.

ak_i = area allocated to crop in zone 1 to 4 during the Kharif season in RBMC, in ha

ar_i = area allocated to crop in zone 1 to 4 during the Rabi season in RBMC, in ha

ap_i = area allocated to crop in zone 1 to 4 during the perennial season in RBMC, in ha

at_i = area allocated to crop in zone 1 to 4 during the biannual season in RBMC, in ha

bk_i = area allocated to crop in zone 5 to 16 during the Kharif season in LBMC, in ha

br_i = area allocated to crop in zone 5 to 16 during the Rabi season in LBMC, in ha

bp_i = area allocated to crop in zone 5 to 16 during the perennial season in LBMC , in ha

bt_i = area allocated to crop in zone 5 to 16 during the biannual season in LBMC , in ha

Computation of Net Benefits

The net benefit of each crop shown in the study area is derived. The necessary information for finding out the net benefits is collected from farmers. The general procedure followed in finding net benefit is as follows;

1. The price and rate of application of seeds per hectares is worked out. This after multiplying gives seed cost.
2. The total fertilizer requirement of each crop per hectare for the crop period is worked out.
3. The cost of cultivation for each crop for the crop period per hectare is worked out.
4. The total cost of pesticides for each crop for the crop period is worked out.
5. The total cost of irrigation water is worked out based on the rates of irrigation water supplied by Karjan reservoir authorities.
6. The sum of seed cost, fertilizer, cultivation, pesticides and irrigation water charges gives total expenditure for each crop per hectare.
7. The yield of each crop per hectares and selling price for each crop is multiplied for deriving gross benefits.
8. After deducting total expenditure from gross benefits, net benefits for each crop per hectare is derived.

It is noted that during the derivation of net benefits it is assumed that the farmer and his family is doing all labour work by their own.

The net benefits are computed using market rates of 2011-12 and it is given in Table 1.

Table 1. Net benefit for crop sown in study area

Sr. No.	Crop	Seed cost (Rs./ha)	Fertilizer (Rs./ha)	Khed (Rs./ha)	Pesticides (Rs./ha)	Irrigation water charges (Rs./ha)	Expenditure (Rs./ha)	Yield (kg/ha)	Selling price (Rs./kg)	Gross benefit (Rs./ha)	NB (Rs./ha)
1	Sugarcane	66666	10000	15000	6000	6465	104131	75000	1.95	146250	42119
2	Banana	57000	626430	15000	400	6465	705295	90000	10	900000	194705
3	Cotton	2325	2500	6250	6250	2990	20315	3125	70	218750	198435
4	Wheat	2250	3625	1250	0	1610	8735	1250	12.5	15625	6890
5	Vegetable	16000	2875	6000	11200	460	36535	1250	32	40000	3465
6	Juvar/Bajra/ Makai	250	500	250	0	460	1460	1250	5	6250	4790
7	Diwela/ Castor	500	2500	6250	1500	2990	13740	3750	50	187500	173760
8	Pulses	750	2500	750	2250	460	6710	2500	30	75000	68290
9	Groundnut	13500	3000	6000	0	460	22960	1750	30	52500	29540

Water Allocation Constraints

The water required for the crop growth in canal has to be met from the reservoir releases, mathematically with the efficiencies of the surface water system the constraint is expressed as:

$$\sum_{t=1}^{12} NIR_{it}\ ak_i \le \eta_s R_t$$

$$\sum_{t=1}^{12} NIR_{it}\ bk_i \le \eta_s L_t$$

$$\sum_{t=1}^{12} NIR_{it}\ ap_i \le \eta_s R_t$$

$$\sum_{t=1}^{12} NIR_{it}\ bp_i \le \eta_s L_t$$

$$\sum_{t=1}^{12} NIR_{it}\ at_i \le \eta_s R_t$$

$$\sum_{t=1}^{12} NIR_{it}\ bt_i \le \eta_s L_t$$

$$\sum_{t=1}^{12} NIR_{it}\ ar_i \le \eta_s R_t$$

$$\sum_{t=1}^{12} NIR_{it}\ br_i \le \eta_s L_t$$

NIR_{it} = Net irrigation requirement of crop "*i*" during month

"*s*" = Efficiency of the surface water system.

The necessary conversions are made in the units of data while using for the study.

The net irrigation requirement considered for the study area is given in Table 2.

Canal Capacity Constraints

Release from the reservoir at any time period of "*t*" must be less than or equal to the carrying capacity of the main canals and is explained as,

$$Rt \le CCt$$

$$lt \le CCt$$

CCt = canal carrying capacity for month "*t*".

The carrying capacity of RBMC considered in study area is maximum at 11.4 mm^3 and minimum at 9 mm^3 and the same for LBMC is maximum at 43.54 mm^3 and minimum at 32.4 mm^3.

Reservoir Storage Capacity Constraints

The reservoir storage in any time period should not be more than the maximum capacity of the reservoir, and should not be less than the minimum storage. The

Table 2. Net irrigation water requirements for each crop in Millimetre.

Sr. No.	Crop	Jan.	Feb.	Mar.	Apr.	May.	June.	July.	Aug.	Sep.	Oct.	Nov.	Dec.
1	Groundnut	0	0	0	0	0	0	0	3	30	31	0	0
2	Sugarcane	115	85	200	165	270	86	0	0	1	0	151	110
3	Banana	105	155	220	255	230	55	0	25	71	66	95	85
4	Jowar/Bajra/Makai	40	80	240	80	0	0	0	3	0	90	70	120
5	Pulses	0	0	0	0	0	0	0	3	23	56	0	0
6	Vegetable	20	20	30	40	115	27	0	2	1	23	45	43
7	Wheat	0	0	0	0	0	3	0	0	47	175	150	100
8	Cotton	150	170	60	0	0	0	0	0	0	0	95	85
9	Diwela/Castor	20	20	30	40	115	27	0	2	1	23	45	43

minimum storage refers to dead storage plus minimum requirement of water for all other purposes.

$S_{min} \leq S_t \leq S_{max}$ Where,

S_{min} = minimum storage capacity observed in 18 years data in mm^3

S_{max} = maximum storage capacity observed in 18 years data in mm^3

S_t = Storage at any time "t"

The data of storage considered in study is given in Table 3.

Table 3. Monthwise minimum and maximum storage for Karjan reservoir project (mm^3)

Month	Min storage in mm^3	Max storage in mm^3
Jan.	110.48	584.00
Feb.	108.89	579.80
March	107.50	570.50
April	96.12	557.10
May	66.31	511.12
June	100.80	454.42
July	107.57	420.20
Aug.	109.93	495.98
Sep.	114.55	625.20
Oct.	114.02	626.48
Nov.	112.66	598.96
Dec.	111.76	587.00

Area Constraints

The total cropped area during any time "t" should not be more than maximum area available for the irrigation.

For RBMC

During the Kharif season

$$\sum_{i=1}^{9} ak_i < a_i$$

$$\sum_{i=1}^{9} ak_i > a_i$$

During the Rabi season

$$\sum_{i=1}^{9} ar_i < a_i$$

$$\sum_{i=1}^{9} ar_i > a_i$$

During the Perennial

$$\sum_{i=1}^{9} ap_i < a_i$$

$$\sum_{i=1}^{9} ap_i > a_i$$

During the Two-Seasonal

$$\sum_{i=1}^{9} at_i < a_i$$

$$\sum_{i=1}^{9} at_i > a_i$$

For LBMC

During the Kharif season

$$\sum_{i=1}^{9} bk_i < b_i$$

$$\sum_{i=1}^{9} bk_i > b_i$$

During the Rabi season

$$\sum_{i=1}^{9} br_i < b_i$$

$$\sum_{i=1}^{9} br_i > b_i$$

During the Perennial season

$$\sum_{i=1}^{9} bp_i < b_i$$

$$\sum_{i=1}^{9} bp_i > b_i$$

During the Two-Seasonal season

$$\sum_{i=1}^{9} bt_i > b_i$$

$$\sum_{i=1}^{9} bt_i < b_i$$

a_i = total cultivable command area for particular crop in RBMC

b_i = total cultivable command area for particular crop in LBMC

The data considered for study is given in Table 4.

Evaporation Constraints

A linear relationship has to be established between volume of evaporation loss in the reservoir and average storage for time "t"

$$E_t = a_t + b_t\,[(S_t + S_{t+1}) / 2]$$

E_t = evaporation loss from the reservoir in mm^3 during the period "t"

S_t = final storage at any time period "t"

S_{t+1} = final storage of reservoir during time period "t"

a_t & b_t = regression coefficients during time period "t"

Using the data of the 18 years, the plot is made for each month between $s\ (S_t + S_{t+1})$ and E_t where S_t stands for initial storage of every month, S_{t+1} is the storage at the end of every respective month and E_t is the evaporation of season "t" stands for time period in month. Monthly linear regression models are developed using 18 years actual data of evaporation. Figures 4 to 15 present linear regression models developed using origin 8.0 software and summary of monthly model is presented in Table 6 with value of statistical parameter "r".

Table 4. Crop area in RBMC and LBMC zone

Crop	Section	Crop notation	Avg. area in ha	Min area in ha	Max area in ha
Groundnut	RBMC	ak1	113.6962	106.78	205.1
Sugarcane	RBMC	ap1	1002.818	137.52	2035.66
Banana	RBMC	ap2	1125.791	50.3	3060.83
Juvar/Bajra/Makai	RBMC	ap3	288.6031	116.67	805.3
Pulses	RBMC	ap4	47.23846	10.8	179.68
Vegetable	RBMC	ap5	10.95615	3.38	50.19
Wheat	RBMC	ar1	44.27538	29.85	138.77
Cotton	RBMC	at1	489.0685	172.83	1494.08
Diwela/Castor	RBMC	at2	14.54615	0.6	113.99
Ground Nut	LBMC	bk1	40.02	5.5	125.87
Sugarcane	LBMC	bp1	3827.016	656.24	7045.8
Banana	LBMC	bp2	422.1154	38.93	1283.92
Juvar/Bajra/Makai	LBMC	bp3	326.6646	15.75	982.36
Pulses	LBMC	bp4	163.7915	21.73	615.67
Vegetable	LBMC	bp5	25.62615	4.92	86.36
Wheat	LBMC	br1	179.7315	84.14	429.42
Cotton	LBMC	bt1	451.4508	25.5	1799.47
Diwela/Castor	LBMC	bt2	47.59	0.4	139.79

The crop area considered in model 2 is given in Table no. 4

Table 5. Planned crop area

Season	% of total area	Area in ha
Perennial	6	2239
Rabi	38	14812
Hot	11	4105
Kharif	50	18661
Two-Seasonal	33	12317
Total	138	51504

Table 6. Statistical summary of 'r' for evaporation

Month	A	b	r
Jan.	1.614	0.003039	0.822
Feb.	1.85	0.00399	0.809
March	2.91	0.00322	0.734
April	2.97	0.00496	0.85
May	1.59	0.0144	0.632
June	2.208	0.0054	0.658
July	-0.0975	0.0111	0.96
Aug.	1.241	0.00548	0.64
Sep.	0.694	0.00751	0.75
Oct.	10.099	-0.0103	0.5892
Nov.	4.33	-0.00136	0.6887
Dec.	1.7945	0.0024	0.6574

Continuity Constraints

The surface water continuity constraint specifies the relationship of month to month reservoir storage and it is mathematically given by

$$S_{t+1} = S_t + I_t - R_t - E_t - Ovf_t$$

$$S_{t+1} = S_t + I_t - L_t - E_t - Ovf_t$$

S_t = storage in mm^3 during month "t"

R_t = releases from the reservoir during month "t" in mm^3

E_t = evaporation Losses during month "t" in mm^3

S_{t+1} = storage in the reservoir during month "t+1" in mm^3

I_t = dependable inflow in mm^3 into reservoir during month "t"

Ovf_t = overflow in mm^3 from the reservoir in the month "t"

Dependable Inflow

The uncertainty in inflows arising due to variations in rainfall is tackled through linear programming. Sixteen years of historical inflow data is used to proposed inflows at 50, 55, 60, 65, 70, 75, 80, 85 and 90% dependability levels. To find dependable inflows for monthly and annually, the Weibull (VenTe Chow) method is used. The computation of dependable flow is appended. Inflows at various dependable flow is tabulated in Table 7 which is considered for present study.

Table 7. Inflow at various dependability for each month

Month	90%	85%	80%	75%	70%	65%	60%	55%	50%
Jan.	0.00	0.055	4.50	0.61	8.36	1.63	0.00	0.06	3.20
Feb.	0.01	0.00	1.40	0.87	0.00	0.00	0.00	3.46	1.17
March	0.00	0.00	1.36	0.00	0.01	0.00	0.00	5.53	0.00
April	0.00	0.00	0.98	0.00	1.19	0.00	0.00	1.10	0.87
May	0.00	0.00	2.26	0.04	0.49	0.00	0.00	0.00	5.45
June	162.97	47.57	87.77	51.78	57.75	24.92	0.55	97.37	289.67
July	75.30	792.27	237.35	167.36	523.27	716.41	90.22	124.81	10.98
Aug.	940.38	915.90	475.51	226.62	371.63	393.20	452.40	85.61	159.58
Sep.	57.38	147.40	39.98	147.11	147.48	240.30	378.37	499.59	308.87
Oct.	32.87	39.90	27.43	25.82	58.78	30.40	74.87	74.76	6.17
Nov.	4.14	7.12	5.91	1.25	6.67	1.09	9.36	11.68	0.75
Dec.	2.67	0.86	1.64	0.00	4.25	0.30	0.41	4.56	4.53

The annual probability distribution was computed for each and every dependable annual inflow value. In the case of monthly probability distribution, the probability distribution was computed for each month and the corresponding dependable inflow values were considered for model. The annual inflow values for different inflow levels are shown in Figure 2. The monthly inflow values for different inflow levels are shown in Figure 3.

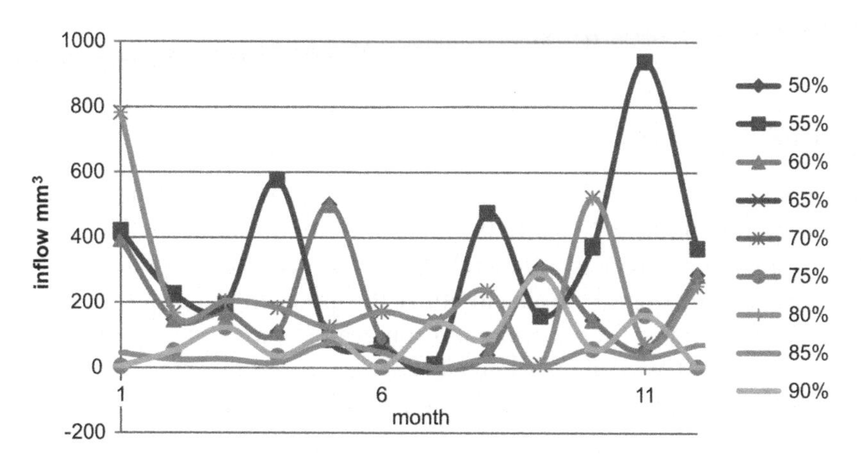

Fig. 2. Annual inflow values for different dependability levels (from July to June)

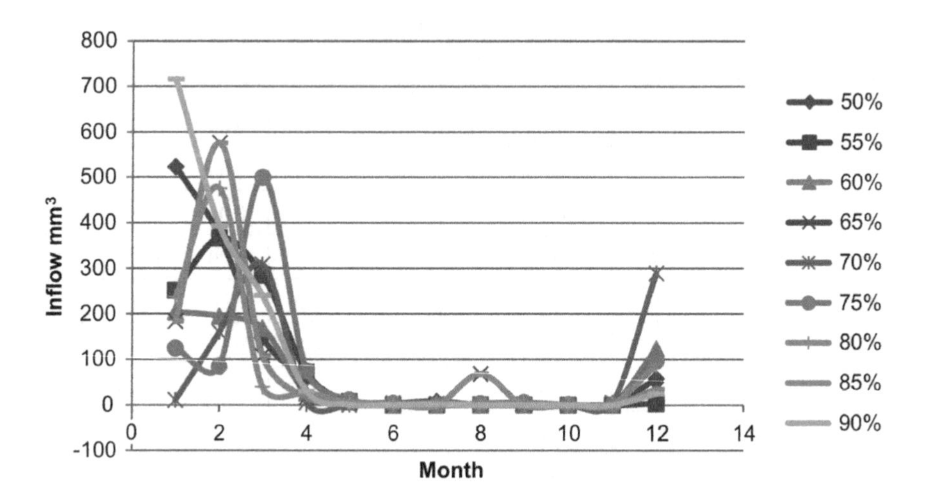

Fig. 3. Monthly inflow values for different dependability levels (from July to June)

Overflow Constraints

If no constraint on overflow is provided, a linear programming model will result in spill even when the reservoir storage is less than capacity, hence the spill constraint developed by Chavez Morales et al. (1987) has been used in this study. The overflow constrain is given by:

$$Ovf_t \geq S_t + I_t - \Sigma(R_t + I_t) - E_t - S_{max}$$
$$t = 1, 2 \ldots 12.$$
$$Ovf_t > 0$$
$$t = 1, 2 \ldots 12$$

SOFTWARE USED

The objective and all the constraints were formulated into a single input file by programming in LINGO 13.0 software (Window based). LINGO stands for Language for Interactive General Optimization and the supplier is LINGO systems. Initially users were able to utilize the modelling language to concisely express models using summation and subscripted variables. In 1994, LINGO became the first modelling language software to be included in a popular management science text. In 1995, the first Windows release of LINGO was made.

RESULT AND ANALYSIS

The results proposed from linear programming model are as follows:

Evaporation

The evaporation resulted from LP Model 1, 2 and 3 at 75% dependable flow is compared with actual average evaporation and it is shown in Table 8 and Figure 4.

Table 8. Monthly evaporation

Month	Actual evaporation observed in 18 years	Model-1(with max crop area)	Model-2(with planned crop area)	Model-3(with average crop area)
Jan.	3.05	2.514	2.5	3.08
Feb.	3.71	2.868	2.85	3.62
March	4.79	3.588	3.58	4.19
April	5.68	3.79	3.79	4.73
May	5.80	3.34	3.34	6.05
June	3.99	2.77	2.77	3.87
July	2.82	2.59	2.59	2.53
Aug.	2.52	3.892	3.89	3.255
Sep.	3.85	5.363	5.363	5.363
Oct.	4.17	3.829	4.84	3.82
Nov.	3.65	3.54	3.8	3.54
Dec.	3.05	3.08	2.59	2.84

To examine the accuracy of the model more closely, the computed evaporation in mm^3 for every month is plotted against the corresponding actual values and it is shown in Figures no. 19, 20 and 21. In these figures the solid line represents the condition of perfect agreement. Very less variation is computed and actual evaporation is observed in case of LP model-3.

The evaporation loss resulting from model 1, 2 and 3 for various dependable inflow levels are shown in Figures 5, 6 and 7 respectively. From these figures the

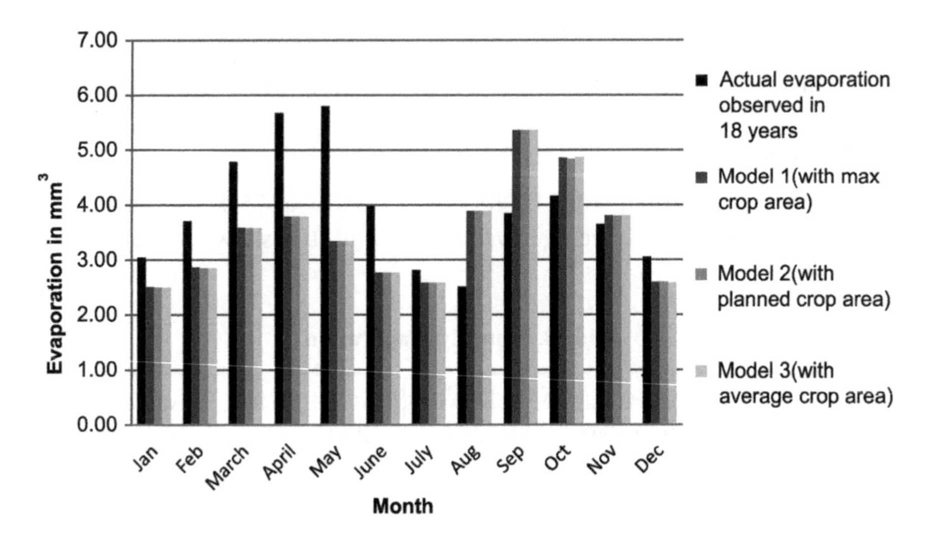

Fig. 4. Evaporation analysis

volume of evaporation loss during kharif season is more because there is more storage in the reservoir and it gradually decreases as the storage decreases, it increases again during summer due to an increase in the evaporation depth.

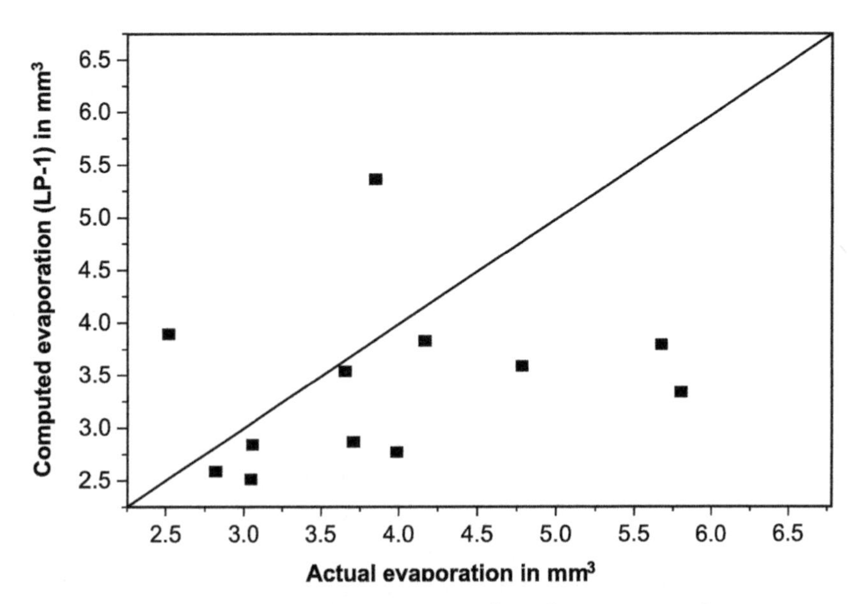

Fig. 5. Computed evaporation plotted against actual evaporation from Model-1

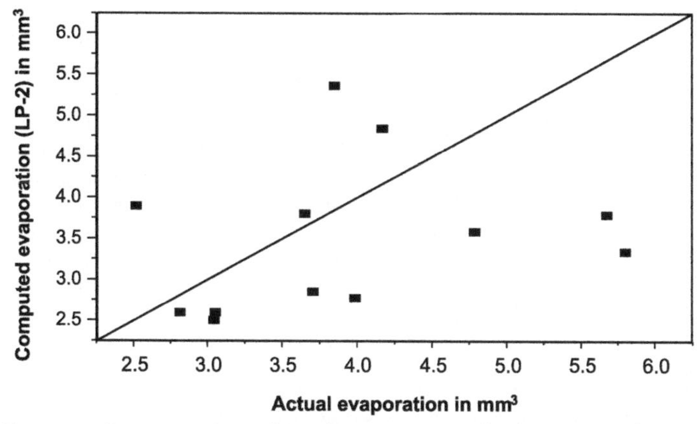

Fig. 6. Computed evaporation plotted against actual evaporation from Model-2

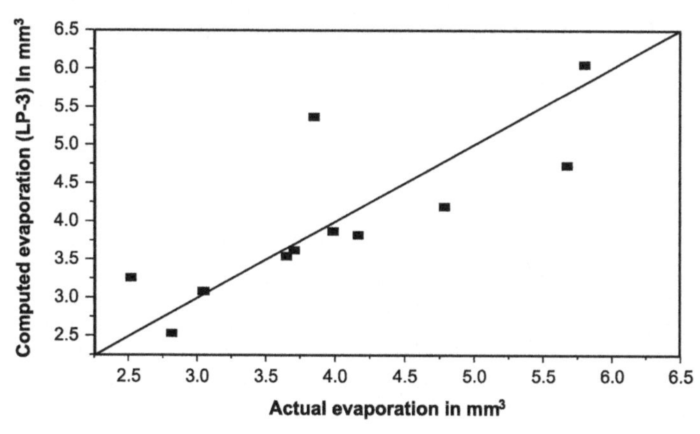

Fig. 7. Computed evaporation plotted against actual evaporation from Model-3

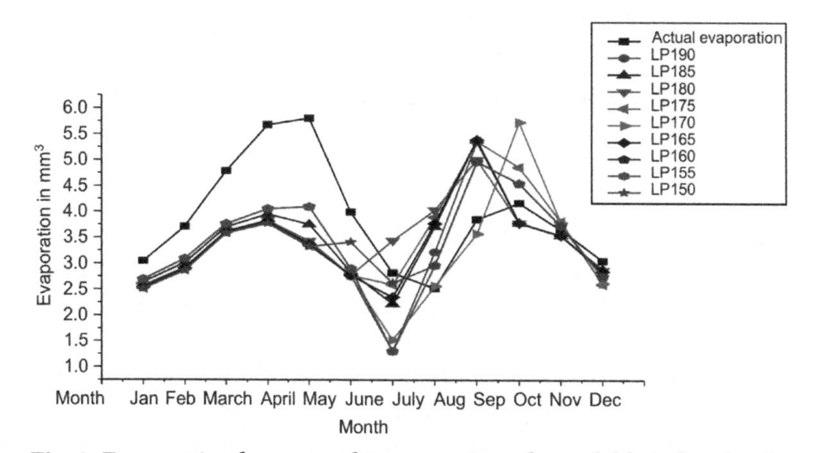

Fig. 8. Evaporation losses resulting at various dependable inflow levels

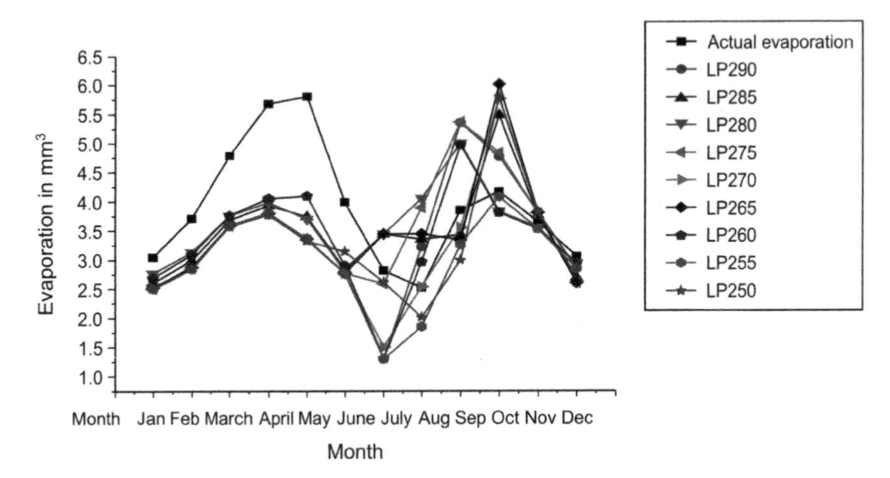

Fig. 9. Evaporation losses resulting at various dependable inflow levels

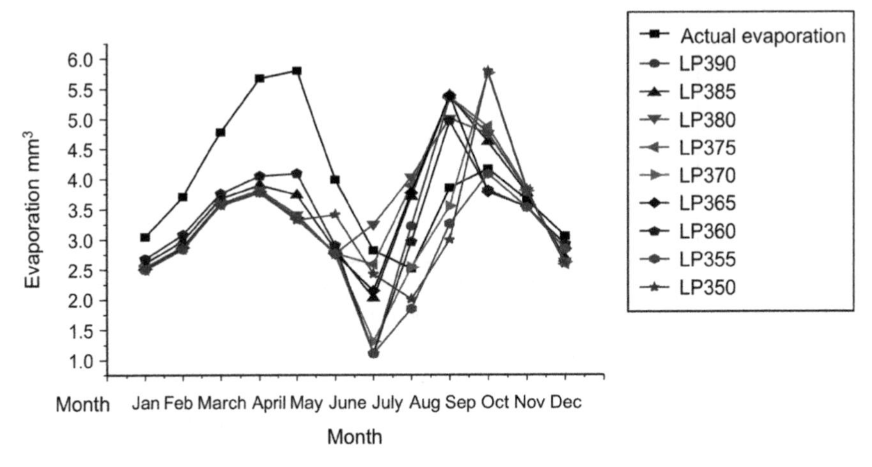

Fig.10. Evaporation losses resulting at various dependable inflow levels

Storage

The storage resulted from LP Model 1, 2 and 3 at 75% dependable flow is compared with actual average storage and it is shown in Table 9 and Fig. 11.

Figure 11 and Table 9 shows that storage reaches to its maximum level in the month of September which is more than 625 mm^3 and less storage reaches to its minimum level in the month of June which is nearly equal to 100 mm^3.

To examine the accuracy of the model more closely, the computed storage in mm^3 for every month is plotted against the corresponding actual values and it is shown in Figs. 8, 9 and 10. In these figures the solid line represents the condition of perfect agreement. Very less variation is computed and actual storage is observed in case of LP Model-2.

Table 9. Monthly storage

Month	Actual storage observed in 18 years	LP 1 (with max crop area)	LP 2 (with planned crop area)	LP 3 (with average crop area)
Jan.	347.24	316.015	313.38	505.58
Feb.	344.345	276.6	273.97	465.59
March	339	233.73	231.12	421.97
April	326.61	187.7	187.49	377.73
May	288.715	143.35	143.28	332.88
June	277.61	100.8	100.8	287.39
July	263.885	107.57	107.57	329.88
Aug.	302.955	342.32	342.32	109.93
Sep.	369.875	625.2	625.2	625.2
Oct.	370.25	618.41	618.41	618.41
Nov.	355.81	598.96	402.35	598.96
Dec.	349.38	552.92	363.057	559.92

The storage loss resulting from model 1, 2 and 3 for various dependable inflow levels are shown in Figs. 12, 13 and 14 respectively. From these figures the volume of storage during kharif season is more because there is more inflow and it gradually decreases as the inflow decreases.

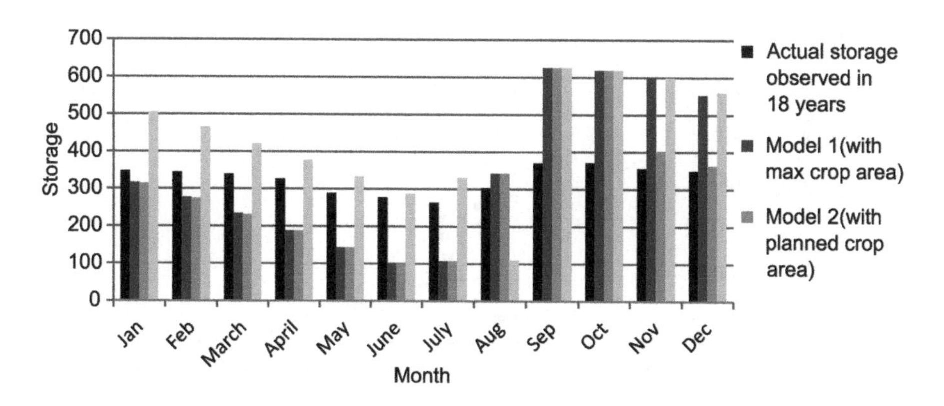

Fig. 11. Storage analysis

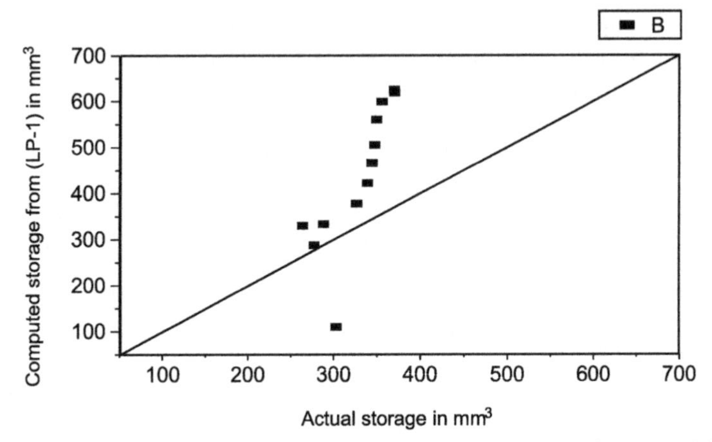

Fig. 12. Computed storage plotted against actual storage from model-1

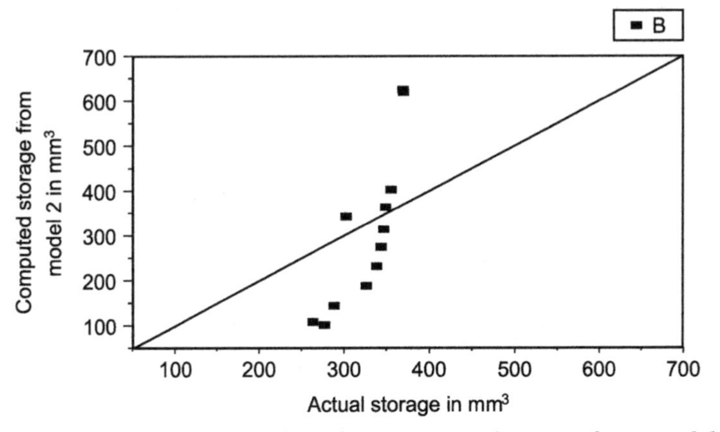

Fig. 13. Computed storage plotted against actual storage from model-2

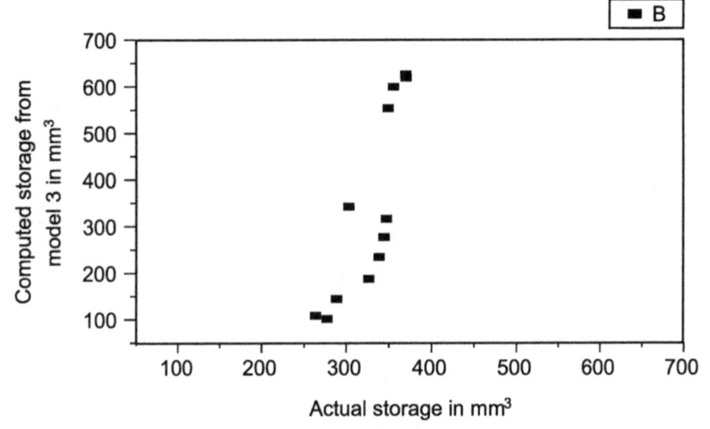

Fig. 14. Computed storage plotted against actual storage from model-3

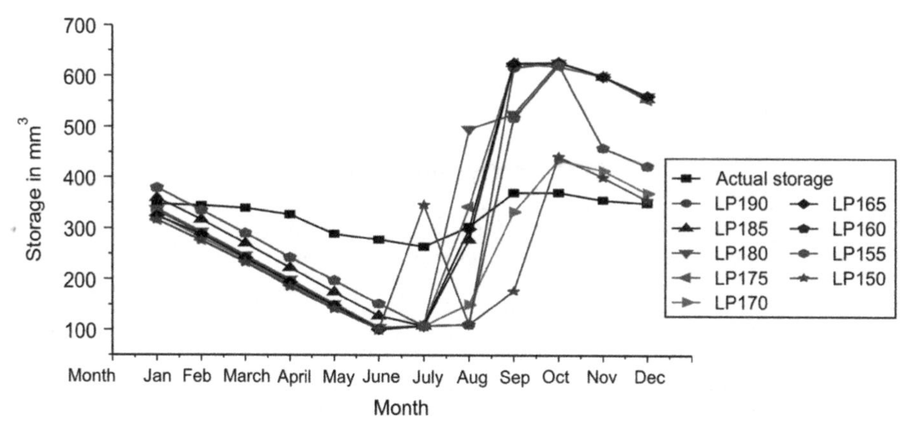

Fig. 15. Storage losses resulting at various dependable inflow levels

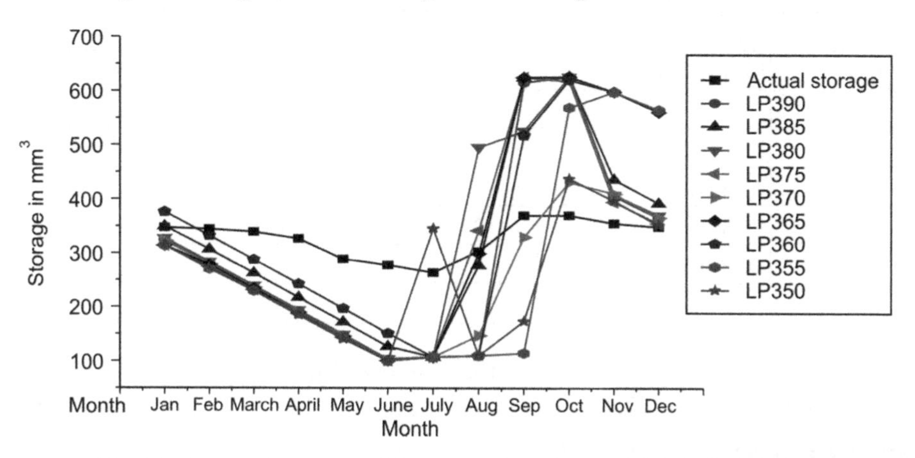

Fig. 16. Storage losses resulting at various dependable inflow levels

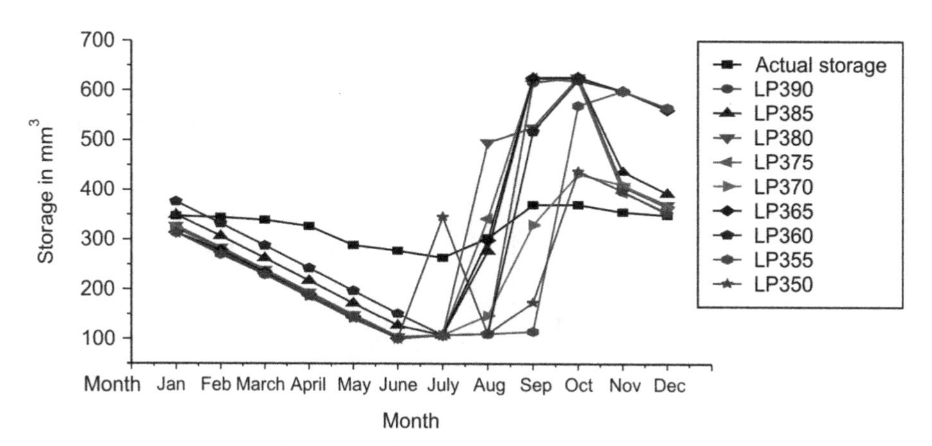

Fig. 17. Storage losses resulting at various dependable inflow levels

Overflow

The overflow resulted from LP Model 1, 2 and 3 at 75% dependable flow is compared with actual average overflow and it is shown in Table 10 and Fig. 18.

Table 10. Overflow analysis

Month	Actual overflow observed in 18 years	LP 1(with max crop area)	LP 2(with planned crop area)	LP 3(with average crop area)
Jan.	0	0	0	0
Feb.	4.610588235	0	0	0
March	0	0	0	0
April	0.281176471	0	0	0
May	6.084705882	0	0	0
June	78.10058824	36.82	36.82	36.82
July	227.1717647	0	0	0
Aug.	248.4658824	147.34	147.34	147.34
Sep.	90.56529412	0	0	0
Oct.	10.86352941	204.51	202.38	201.3
Nov.	0.237058824	0	0	0
Dec.	0	0	0	0

Figure 18 and Table 10 shows that the overflow reaches to its maximum value in the month of October which is more than 200 mm^3 and reaches to its minimum value in the month of June which is more than 35 mm^3.

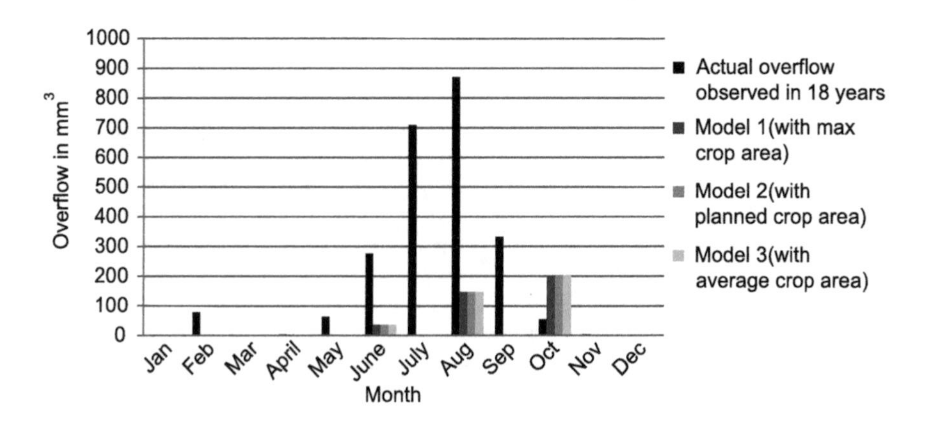

Fig. 18. Over flow analysis

To examine the accuracy of the model more closely, the computed overflow in mm^3 for every month is plotted against the corresponding actual values and it is shown in Figs. 15, 16 and 17. In these figures the solid line represents the condition of perfect agreement. Very less variation in computed and actual overflow is observed in case of LP model-3.

The overflow resulting from model 1, 2 and 3 for various dependable inflow levels are shown in Figs. 19, 20 and 21 respectively. From these figures the volume of overflow during kharif season is more because there is more inflow and it gradually increases as the inflow increases.

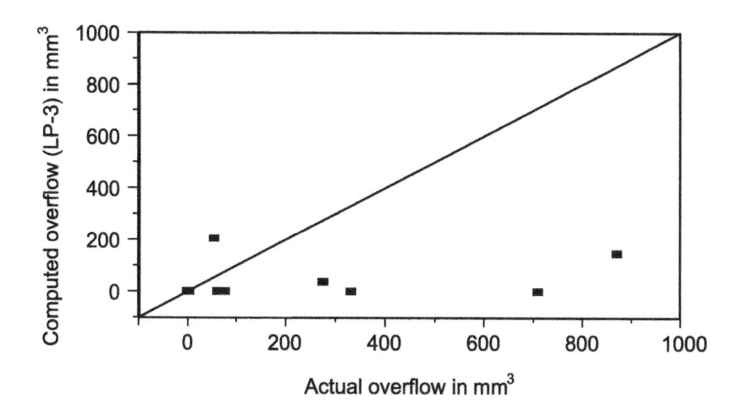

Fig. 19. Computed overflow plotted against actual overflow from model-1

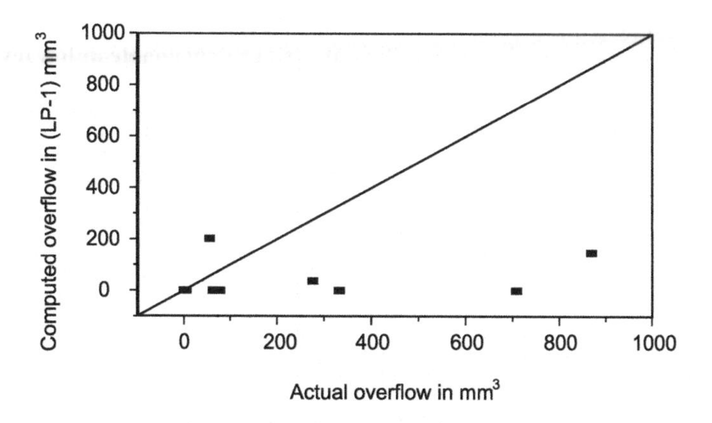

Fig. 20. Computed overflow plotted against actual overflow from model-2

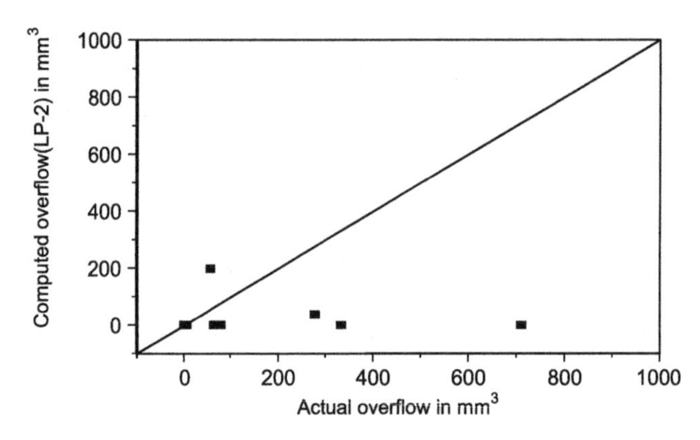

Fig. 21. Computed overflow plotted against actual overflow from model-3

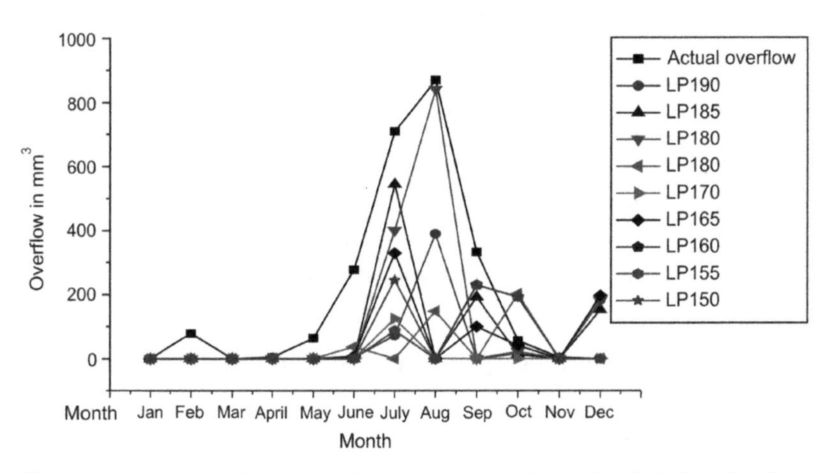

Fig. 22. Evaporation losses resulting at various dependable inflow levels

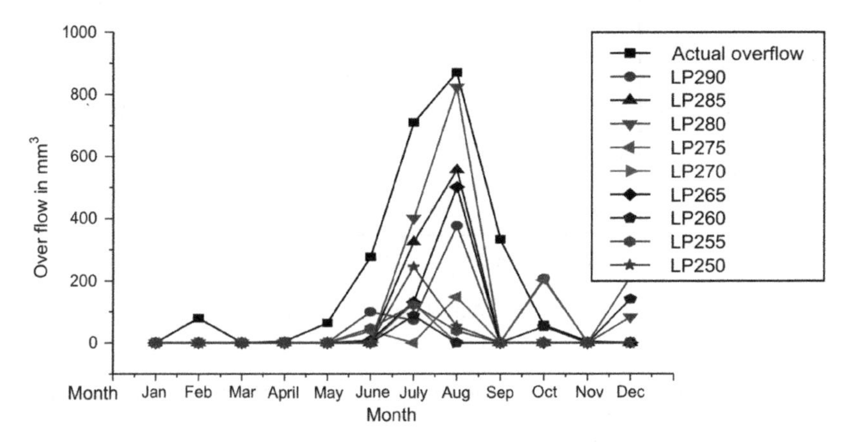

Fig. 23. Evaporation losses resulting at various dependable inflow levels

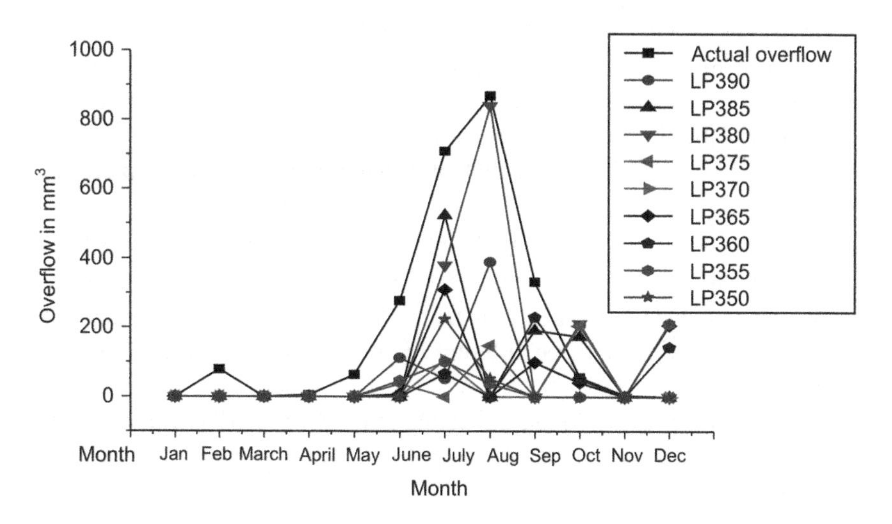

Fig. 24. Evaporation losses resulting at various dependable inflow levels

Crop Area

Crop Area for Right Bank Main Canal (RBMC)

The crop area for each crop resulted from LP model-1, 2 and 3 at 75% dependable flow is compared and it is shown in Fig. 25.

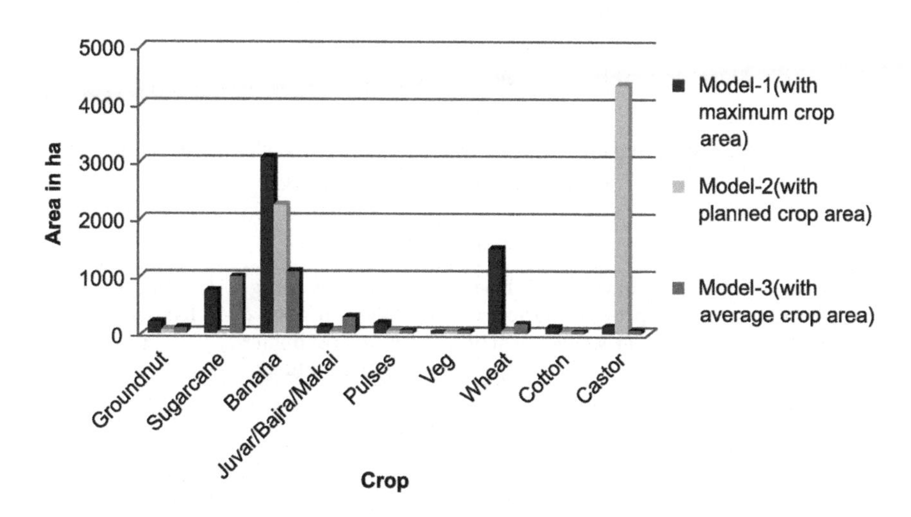

Fig. 25. Crop area analysis of RBMC

From Fig. 25 following findings can be summarized.

1. From model-1 it is seen that banana, wheat and sugarcane should be sown in more area, while groundnut, castor, pulses and juwar/bajra/makai should be sown in less area.
2. From model-2 it is seen that only banana and castor crop should be sown.
3. From model-3 it is seen that sugarcane and banana should be sown in more area, while pulses, cotton, castor should be sown in less area.

The crop area sown in each year from 1999 to 2011 is also compared with crop area computed from three different LP models and it is shown in following figures.

The year wise comparison of actual and computed cropping area for the year 1999 is shown in Fig. 26 from this study following findings can be summarized.

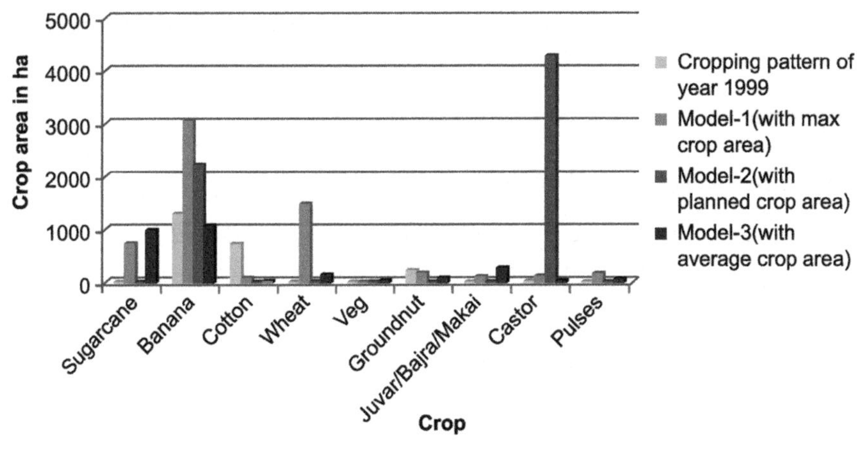

Fig. 26. Comparison of Crop Area Sown in 1999 with Crop Area Computed from Model for RBMC

Year - 1999

Sr. No.	Description of model	Name of crops having proposed crop area nearly equal to actual	Name of crops having proposed crop area higher than actual	Name of crops having proposed crop area less than actual
1.	LP-1	Groundnut	Sugarcane, banana and wheat crops	Cotton
2.	LP-2	Jowar/bajara/makai, vegetables and pulses	Banana and castor	Cotton and groundnut
3.	LP-3	Castor, pulses and vegetables	Sugarcane, jowar/bajra/makai and wheat	Banana and groundnut

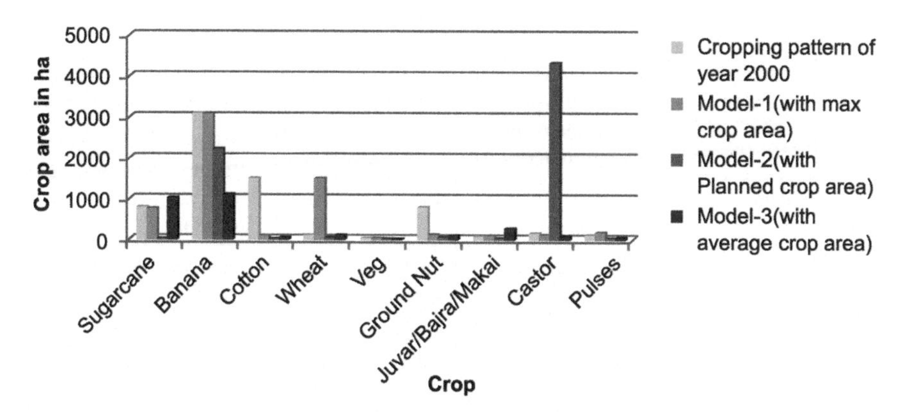

Fig. 27. Comparison of crop area sown in 2000 with crop area computed from model for RBMC

Year-2000

Sr. No.	Description of model	Name of crops having proposed crop area nearly equal to actual	Name of crops having proposed crop area higher than actual	Name of crops having proposed crop area less than actual
1.	LP-1	Sugarcane, pulses, jowar/bajara/makai and banana.	Wheat	Cotton and castor
2.	LP-2	Jowar/bajara/makai, vegetables and pulses	Castor	Cotton, sugarcane, banana, wheat and groundnut
3.	LP-3	Sugarcane, wheat and vegetables	Jowar/bajara/makai	Banana and groundnut

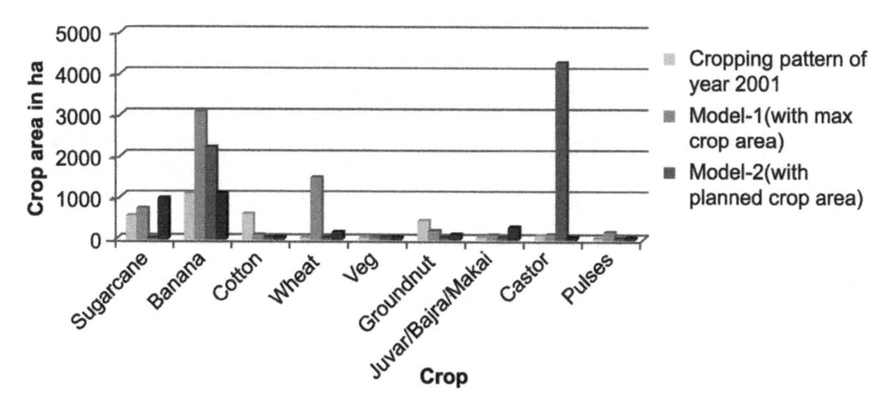

Fig. 28. Comparison of crop area sown in 2001 with crop area computed from model for RBMC

Year-2001

Sr. No.	Description of model	Name of crops having proposed crop area nearly equal to actual	Name of crops having proposed crop area higher than actual	Name of crops having proposed crop area less than actual
1.	LP-1	Jowar/bajara/makai and pulses	Banana and wheat	Sugarcane, castor and groundnut
2.	LP-2	Vegetables and jowar/bajara/makai	Castor and banana	Sugarcane, cotton and wheat
3.	LP-3	Pulses and vegetables	Banana and jowar/bajara/makai	Sugarcane, cotton, castor and groundnut

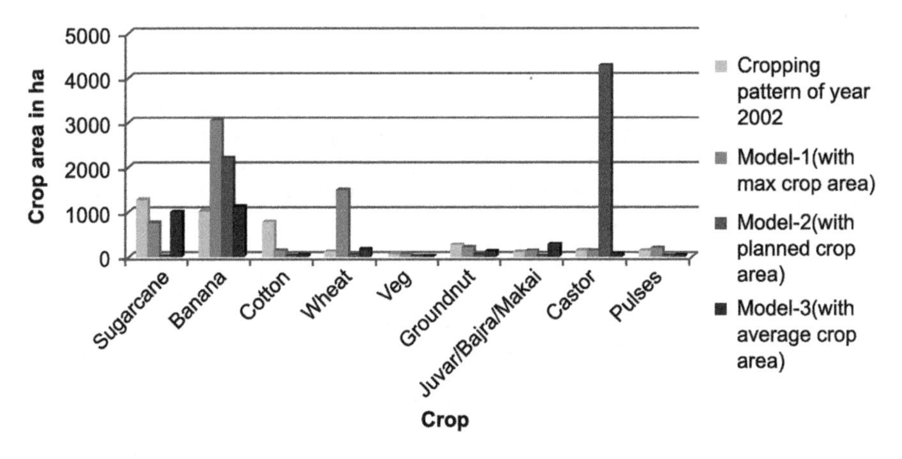

Fig. 29. Comparison of crop area sown in 2002 with crop area computed from Model for RBMC

Year-2002

Sr. No.	Description of model	Name of crops having proposed crop area nearly equal to actual	Name of crops having proposed crop area higher than actual	Name of crops having proposed crop area less than actual
1.	LP-1	Jowar/bajara/makai and vegetables	Banana and wheat	Sugarcane and groundnut
2.	LP-2	Vegetables, pulses and jowar/bajara/makai	Castor and banana	Sugarcane, cotton and groundnut
3.	LP-3	Wheat, pulses and vegetables	Banana and jowar/bajara/makai	Sugarcane, cotton, castor and groundnut

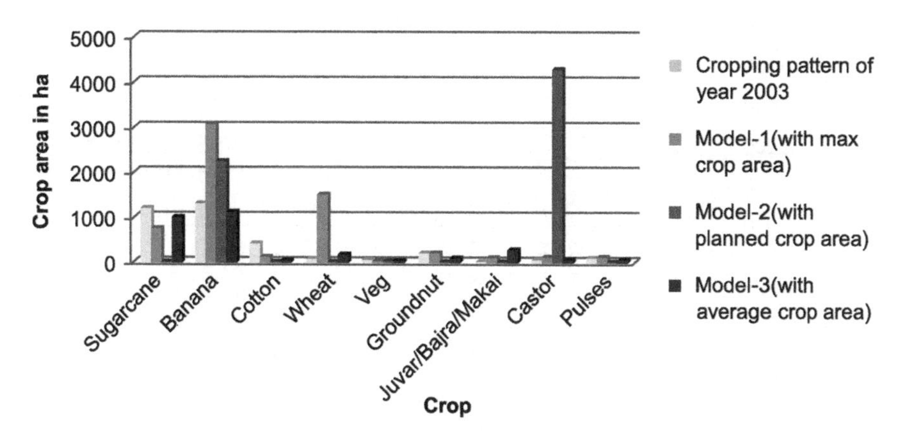

Fig. 30. Comparison of crop area sown in 2003 with crop area computed from Model for RBMC

Year-2003

Sr. No.	Description of model	Name of crops having proposed crop area nearly equal to actual	Name of crops having proposed crop area higher than actual	Name of crops having proposed crop area less than actual
1.	LP-1	Jowar/bajara/makai and vegetables	Banana and wheat	Sugarcane and castor
2.	LP-2	Vegetables, pulses and jowar/bajara/makai	Castor and banana	Sugarcane, cotton and groundnut
3.	LP-3	Wheat, pulses and vegetables	Banana and jowar/bajara/makai	Sugarcane, cotton, castor and groundnut

Note: The above calculations and graphs are only for RBMC and five years but it will be more than five years and also for LBMC.

TOTAL NET BENEFIT

The net benefit achieved from LP model at 75% dependable flow for all three models is compared with the total net benefit achieved in the last 11 years. The comparison is shown in Figure 31.

From the following study following findings can be summarized.

1. The maximum net benefit can be derived if proposed planned cropping pattern is implemented.
2. It is recommended to increase sowing area to increase the net benefits.

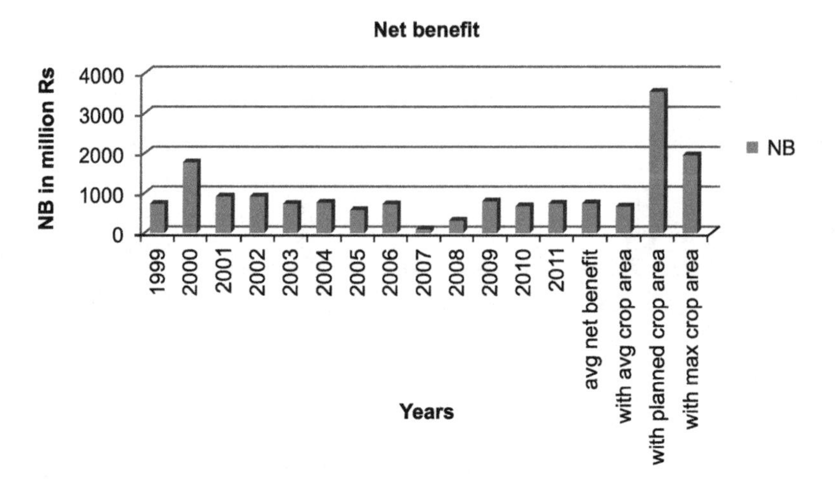

Fig. 31. Net benefit

3. When net benefits derived considering thirteen average area of cropping pattern the net benefits derived obtained from LP model is in close association with the actual net benefits.

CONCLUSION

A linear programming model were developed and applied to Karjan Reservoir Project, Karjan Dam, India to arrive at optimal cropping patterns and reservoir operating strategies, by incorporating the stochastic nature of inflows into its equivalent deterministic inflow into its deterministic form.

These equivalent deterministic inflow values were estimated from monthly probability distributions. Results from three models, one model with planned sowing area, maximum sowing area and average sowing area of last 11 years were compared with respect to net benefit, total cropped area, optimal cropping pattern, evaporation losses and initial storages.

The model with planned cropping area resulted in higher net benefit and higher cropped area than other two models.

The total net benefits resulted from the average cropping area based model when compared with actual net benefits of last 11 years cropping pattern of study area at 75% dependable flow having less variation when compared with other two models.

The total crop area for all the models does not change to the probability distribution of inflow, whereas the evaporation, storage and overflow shows a similar trend was found for evaporation losses and initial storage.

From this study, it can be concluded that consideration of planned cropping pattern may result in exaggerated picture, in terms of net benefits and total cropped area though desirable.

It is recommended to the authority to increase cropping area in the command area of Karjan Reservoir Project to have maximum benefits from the project.

This was also observed from evaporation losses and storage curves. The similar trend was observed for evaporation losses, overflow and storage in case of all the models.

However the evaporation losses and storage values are higher from July to December and lower than actual values from January to June at different level of dependable flow.

The overflow curves follow similar trend in all the models.

REFERENCES

Adeyemo, Josiah and Otieno, Fred. (2009). "Optimizing Planting Areas using Differential Evolution (DE) and Linear Programming (LP)". *International Journal of Physical Sciences,* 4(4), pp. 212-220, April.

Anwar, Arif, A. and Clarke, Derek. (2001). "Irrigation Scheduling using Mixed Integer Linear Programming". *Journal of Irrigation and Drainage Engineering,* ASCE, 127(2), March/April, pp. 63-69.

Azis Hoesein, Abdul and Montarcih Limantara, Lily. (2010). Linear Programming Model for Optimization of Water Irrigation Area at Jatimlerek of East Java, *International Journal of Academic Research,* 2(6). Part 1.

Chavez-Morales, J., Marino, M.A. and Holzapfel, E.A. (1987). Planning Model of Irrigation District. *Journal of Irrigation and Drainage Engineering,* 113(4), pp. 549-564.

Das, Amlan And Datta, Bithin. (2010). "Application of Optimization Techniques in Groundwater Quantity and Quality Management". *Sadhana,* 26, Part 4, pp. 293-316.

English, M.J., Solomon, K.H. and Hoffman, G.J. (2002). "A Paradigm Shift in Irrigation Management". *Journal of Irrigation and Drainage Engineering,* 128(5), pp. 267-277.

Frizzonel, J.A., Coelho, R.D., Dourado-Neto, D. and Soliant, R. (1997). "Linear Programming Model to Optimize the Water Resource Use in Irrigation Projects: An Application to the Senator Nilo Coelho Project". *Sci.Agric,Piracicaba,* 54, pp. 136-148.

Futagami, T., Tamai, N. and Yatsuzuka, M. (1976). "FEM Coupled with LP for Water Pollution Control". *J. Hydraul. Div., Am. Soc. Civ. Eng.,* 102(HY7), pp. 881-897.

Jonoski, Andreja, Zhou, Yangxiao and Nonner, Jan. (1997). "Model-Aided Design and Optimization of Artificial Recharge-Pumping Systems". *Hydrological Sciences-Journal-Des Sciences Hydrology,* 42(6), pp. 937-954.

Jyothiprakash, V. (2000). Optimal Conjunctive Use Operation in Irrigation Systems. Ph.D. Thesis, Indian Institute of Technology.

___________. (2007). "Optimal Crop Planning using Linear Programming Model – A Case Study", Lecture Notes on Soft Computing Techniques in Water Resources Management, *QIP Short Term Course*, Nov, pp. 73-82.

Loucks, D.P., Kindler, J. and Fedra, K. (1985). "Interactive Water Resources Modeling and Model Use: An Overview". *Water Resources Research*, 21(2), pp. 95-102.

Loucks, D.P., Stedinger, J.R. and Haith, D.A. (1981). *Water Resources Systems Planning and Analysis.* Prentice-Hall, Inc.: Englewood Cliffs, NJ.

Montarcih Limantara, Lily. (2010). "Possible Climate Change Effect on Water Irrigation at Golek, Malang, Indonesia". *Journal of Economics and Engineering*, 1(3), pp. 15-17, ISSN: 2078-0346.

Tzimopoulos, C. and Ginidi, P. (2005). "Optimized Aquifer Management, Using Linear Programming. An Application to the Agia Varvara Aquifer, Drama, Greece ,*Global Nest Journal*, 7(3), pp. 395-404.

Weragala, D.K. Neelanga. (2010). Water Allocation Challenges in Rural River Basins: A Case Study from the Walawe River Basin, Sri Lanka, *All Graduate Theses And Dissertations. Paper 589*. Network, Ph.D. thesis.

SUBJECT INDEX

AUTHOR INDEX